Become an Inventor

Idea-Generating and Problem-Solving Techniques with Elements of TRIZ, SIT, SCAMPER, and More

Adam Adrian Brostow

Become an Inventor

Idea-Generating Techniques and Problem-Solving Techniques with Elements of TRIZ, SIT, SCAMPER, and More

ISBN-13: 978-1508936831
ISBN-10: 1508936838

Cover design and photos by Adam A. Brostow
Figures by Adam A. Brostow, except 1.1, 1.2, 1.3, 10.2, 16.1 (part), 16.2 (part), 16.4.

Contents

INTRODUCTION

> Give instruction to a wise man, and he will be yet wiser;
> teach a just man, and he will increase in learning.
> —Proverbs 9:9 (KJV)

> If there's a book that you want to read, but it hasn't
> been written yet, then you must write it.
> —Toni Morrison

We live in the era of globalization. Routine work gets outsourced to the lowest bidder. If we want to keep our jobs, we have to become more and more creative, as an individual's creativity cannot be outsourced.

In addition, we live in the age of ideas. Machines, devices, and brick-and-mortar factories become secondary to ideas and intellectual property.

There are various books written about creativity tools and idea-generating techniques. Many of them are quite good. However, I would like to see a concise book, about one hundred pages long, that describes various idea-generating techniques and creativity tools and the relationship between them, and illustrates them with simple examples and drawings, then progresses to more complex examples. I prefer pictures and cartoons to detailed descriptions (a picture is worth a thousand words!). I did not find one like this so I undertook to write such a book.

I worked as an engineer for four companies for over twenty years. I hold twelve US patents, mostly as a coinventor (it is usually a team effort), with several international equivalent and nonequivalent patents. I am not as good at coming up with an original idea as I am in building on an idea and coming up with a different variation. Sharing my professional time between creating inventions and other intellectual property and routine engineering designs keeps me more grounded compared to somebody working full time in research and development.

I also work with university engineering students, helping with senior design projects, plus teaching the art of invention to schoolchildren at a local Inventor's Lab. Students and, especially, children are often more creative than seasoned engineers and scientists as the younger ones are not afraid to explore new venues. I learn as much from them as they learn from me.

My other experience includes Engineers Without Borders' projects in Africa and Central America. Such projects require a lot of improvisation with minimum resources.

I am also an exhibited painter and a published author, and I find a strong connection between creativity in different disciplines.

Some fifteen years ago I stumbled into TRIZ. TRIZ is a Russian acronym for *теория решения изобретательских задач*, which translates as "the theory of inventive problem solving." It was developed in the former Soviet Union by Genrich Altshuller, starting in 1946. There are several books written about TRIZ, so I do not want to replicate that here. To me TRIZ mostly provides a checklist of things to look at. Many scientists, engineers, and inventors use TRIZ–like techniques intuitively, without knowing it or applying any particular technique. Others can benefit from a more structured approach.

Many creativity tools and brainstorming experts work as full-time consultants. I am an actual inventor, applying different techniques to real-life problems, producing patents, designing new processes and new pieces of equipment, and improving existing designs. Like Altshuller (and Albert Einstein), I spent part of my professional life analyzing patents.

I was once asked to lead the TRIZ community of interest at work. Over several years I headed and coheaded several brainstorming and problem-solving sessions using TRIZ and similar methodology. We generated many excellent ideas, and some were implemented or patented.

The TRIZ methodology includes the 40 Inventive Principles. It also includes the concepts of Ideal Final Result, Contradictions, and the four Separation principles. Some of the 40 Inventive Principles are quite general and can be used to solve a variety of problems. Others are quite specific and were originally developed to solve engineering problems, based mostly on patented technologies used in mechanical and civil engineering in the former Soviet Union during the 1950s and 1960s.

One of the early idea-provoking techniques was called SCAMPER. It was developed in the 1950s and '60s in the States by people like Alex Osborn and Bob Eberle. It includes seven techniques that form the SCAMPER acronym: Substitution, Combination, Adaptation, Modification, Put to Other Purposes, Elimination, and Reverse/Rearrange. SCAMPER was developed in America independent of TRIZ.

Many of the SCAMPER techniques are quite similar to TRIZ. For a list of the forty TRIZ principles, go to the section entitled 40 Inventive Principles, SIT and SCAMPER found herein. For instance SCAMPER's Elimination corresponds to TRIZ's Subtraction. I think this is an example of convergent evolution. The contribution of people like Altshuller and Osborn were not to the development of specific techniques used for millennia. Their contribution was to the classification and systematization of those often quite obvious techniques.

Yet another powerful technique developed relatively recently in Israel is SIT—Structured Inventive Thinking. It has only five thought-provoking techniques: Subtraction, Unification, Multiplication, Division, and Attribute Dependency. As one can see, most of the SIT techniques are named after mathematical functions. SIT also uses the Closed World concept. SIT was developed as a modification and simplification of TRIZ.

The techniques mentioned above are often used somewhat dogmatically. For example, some TRIZ purists insist that one should always identify contradictions. SIT postulates that one cannot go outside a certain domain in seeking a solution. Those approaches work well for a number of problems, but a more flexible approach can often yield better results.

To truly evaluate the efficacy of TRIZ, SIT, SCAMPER, and other creativity tools, one would have to do a double-blind test. One could divide participants into two random groups and have one group solve a problem using a particular tool and have another group solve the same problem using another technique or no technique at all. I think putting smart people in one room is already a great way to solve a problem and generate ideas. TRIZ, or another methodology, provides some structure for them.

Some creative problem solvers, when presented with TRIZ and similar techniques, roll their eyes and say there is nothing new or original about it. Others are big enthusiasts. Yet others make a living selling those techniques so they have an agenda. I personally find such techniques useful, whether they improve my creativity by 10 % or by 100 %.

Creativity tools and idea-generating techniques evolve in two independent directions, towards simplicity and towards increased complexity, often claiming being able to solve any problem. I tend to choose simplicity. I also tend to use techniques that are common to different methods developed independently, such as to TRIZ and SCAMPER. This I think validates them and eliminates the bias such as the bias in Soviet technology of the past century.

I find some of the TRIZ principles too specific and not very useful. Conversely I find some of the techniques of simpler systems too general, where some of the techniques can be seemingly applied to come up with any solution. I simply use the techniques I like, maybe fifteen of them. For this reason I titled my book *Elements of TRIZ, SIT, SCAMPER*. For more conventional approaches to the above-mentioned creativity tools, I refer the reader to the Literature section herein.

This book should be useful to engineers, scientists, students, individuals working with intellectual property, as well as everyday problem solvers and tinkerers. Unlike many books on this subject, this book uses well-known examples in addition to my original examples from my own diverse experiences. The complex examples are often taken from chemical or process engineering and may be a bit intimidating to a nonexpert. But I hope nonengineers can learn from the simple examples and apply the concepts to complex problems in their respective disciplines while complex examples prove that those techniques can be applied to complex and non-trivial problems.

Many concepts can be used for problem solving outside of science or engineering, in writing, art, music, business. In addition to presenting known concepts and illustrating them with old and new examples, I introduce a number of my own original ideas. They include concepts like The Enabler and Taking It to the Extreme in problem solving. I often describe my own use and interpretation of various principles instead of describing the conventional use, with the name of the principle a mere suggestion of how it can be applied.

GYRO GEARLOOSE

Carl Barks was a cartoon artist who worked for Disney. He is probably the only cartoon artist quoted as prior art in an actual patent application.

In a 1949 cartoon *The Sunken Yacht* Donald Duck and his nephews salvage a ship from the bottom of the ocean by filling the hull with ping-pong balls until it simply floats to the surface (Fig. 1.1):

Fig. 1.1 ©1949 Disney

A Danish inventor was refused a Dutch patent for a similar invention involving raising a sunken ship by filling the hull with buoyant bodies fed through a tube (Fig. 1.2).

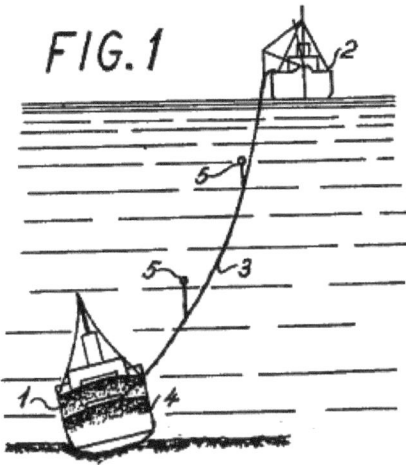

FIG. 1

Fig. 1.2

Gyro Gealoose, an anthropomorphic chicken inventor, was created by Carl Barks as part of Donald Duck's universe. He lives in Duckburg.

In a 1956 cartoon *Grandma's Present*, Gyro improves a farm by eliminating a cow and food crops. He observes that a milk-producing cow eats hay and drinks water. He creates a machine that converts hay and water directly to milk. He then takes his invention further. He observes that grass grows on soil. He modifies the machine to convert dirt directly to milk, fruits, vegetables, and meat (Fig. 1.3).

Gyro implicitly uses TRIZ Extraction principle, SIT's Subtraction, or SCAMPER's Elimination. Please notice that in 1956 TRIZ was in its infancy in the Soviet Union, and SIT and SCAMPER did not exist.

Even today Gyro's ideas make sense. Milk-like drinks can be made directly from plants, for example from soy. Meat can be grown in a laboratory using basic nutrients.

In another cartoon, *Crow Foe*, Gyro implicitly uses a SIT postulate that a solution should not only solve the problem but also reverse the negative outcome. Gyro invents an improved scarecrow that not only scares away the crows but also makes them return the corn they previously stole.

Fig. 1.3 ©1956 Disney

SUBTRACTION (EXTRACTION, ELIMINATION)

La perfection est atteinte non quand il ne reste rien à ajouter, mais quand il ne reste rien à enlever. (You know you've achieved perfection in design, not when you have nothing more to add but when you have nothing more to take away.)
—Antoine de Saint-Exupery

David was already in the stone. I just took away what wasn't David.
—Michelangelo

This is probably the most powerful technique, if ostensibly obvious. The principle is called Extraction in TRIZ (Principle 2), Object Removal or Subtraction in SIT, and Elimination in SCAMPER. TRIZ teaches to extract the disturbing part or only the necessary part. Extracting the disturbing part often means separating it from the rest of the system. However, the part may remain connected to the system. Extracting the necessary part means eliminating all the remaining parts, and identifying and retaining the essential part(s). SIT calls for removing an element and replacing it with an existing element (see section on Unification). SCAMPER seeks to remove various elements and see what happens.

Subtraction can be seen in nature. Snakes lost their legs in an environment where they did not need it. We lost most of our body hair. We lost opposing toes on our feet as we adapted to walking.

Here is my approach. We can remove parts of a system/device (or steps of a process) that serve no purpose and may only be there for historical reasons. This is called trimming. We can remove parts for Partial Benefit (a variation of TRIZ Principle 16, Partial or Excessive Action). We can remove parts so that other existing parts can take up the same function. Finally we can remove parts to come up with a new invention.

It is always good to examine an object or a process we plan to improve, divide it into parts, and try to remove each part in turn. Sometimes parts can be removed without changing the functionality. Sometimes another existing part takes up the function of the one that was removed. This part may be adopted with or without modifications. This is called Functional Analysis.

Here is a trivial example. I have a folding reading lamp that does not have an on/off switch. It turns on when unfolded. The switch has been eliminated as unnecessary.

A typical car has a fuel pump and a lubricating system. The old East German Trabant car had no fuel pump and no oil pump. The fuel tank was located above the engine so the fuel entered the engine by gravity. It had a two-stroke engine that used a mixture of gasoline and oil. Of course it was not very efficient or environmentally friendly. However, it was pretty reliable.

There is a saying about knitting with just one needle. The idea is not that silly. After a small modification we have crochet!

Let us see how we can use Subtraction to come up with a new invention. In this case, removing an object may lead to a new application.

Let us look at some trivial examples. Consider a bicycle (Fig. 2.1).

Fig. 2.1

What if we remove one wheel? We have a unicycle, a niche application. What if we remove pedals and the saddle? We have a kick scooter, an example of a new invention.

I once asked a group of ten- to eleven-year-olds from the Inventor's Lab to remove parts from a motorcycle. None of them had ever heard of Dean Kamen's invention called the Segway. After removing the saddle and with some modifications, they came up with something very similar (Fig. 2.2).

Fig. 2.2

What if we eliminate the engine and the propeller from an airplane? We have a glider (Fig. 2.3).

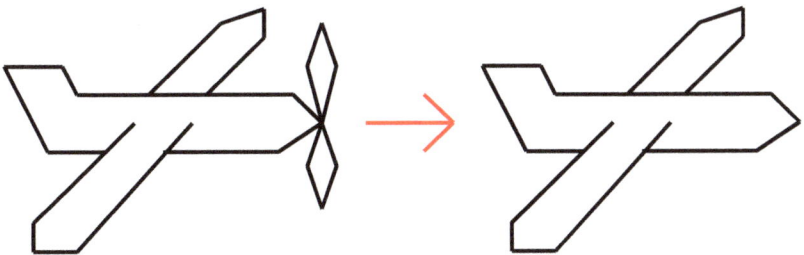

Fig. 2.3

An autogiro is an aircraft with a top propeller usually disconnected from the engine. This aircraft needs a shorter runway to land and to take off.

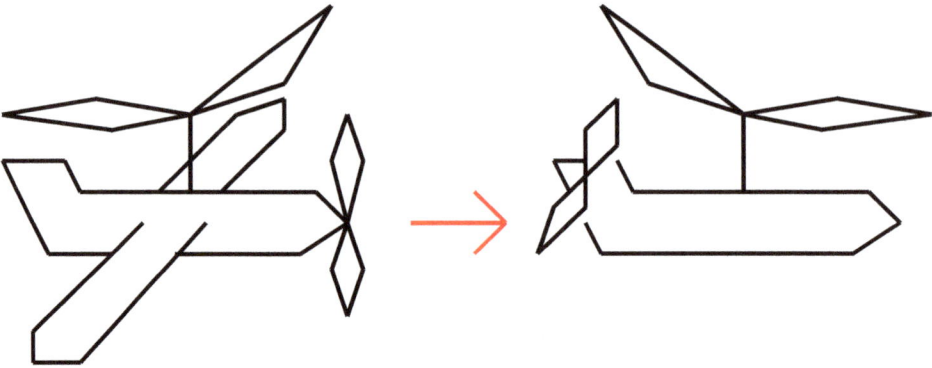

Fig. 2.4

Let us remove the wings. After some modifications (top propeller connected to the engine, smaller stabilizer propeller) we have a helicopter (Fig. 2.4). The first helicopters appeared thirteen years after autogiros. One could have anticipated this design by using the Subtraction principle or simply paying attention to Leonardo da Vinci's drawings.

Here are more Subtraction examples. Let us start with a car. Let us remove the roof – we have a cabriolet. Let us remove the engine—we have a trailer. Let us remove wheels and the engine—we have a simulator. Let us remove the steering from a previously mentioned kick scooter—we have a skateboard. Let us remove wheels from a skateboard—we have a surfboard or a snowboard.

One does not have to be an expert to use Subtraction to come up with ideas for new inventions.

Let us move to more complex examples. Here is a typical distillation column (Fig. 2.5).

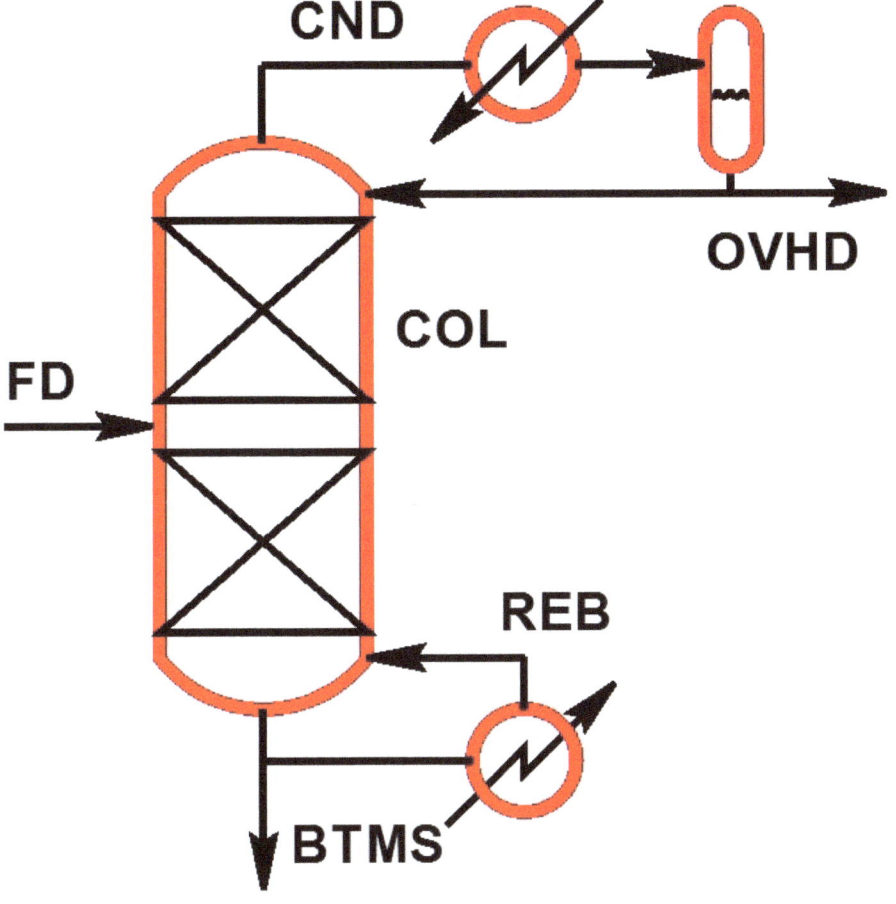

Fig. 2.5

Feed (FD) is separated into the lighter overhead product (OVHD) and the heavier bottoms product (BTMS). The column (COL) had a condenser (CND) on top and a reboiler (REB) at the bottom. The condenser produces liquid flowing down the column (reflux); the reboiler produces vapor flowing up the column. What if we remove the reboiler, a unit that typically requires a heat exchanger and a heating utility such as steam? We now have a rectifier (Fig. 2.6).

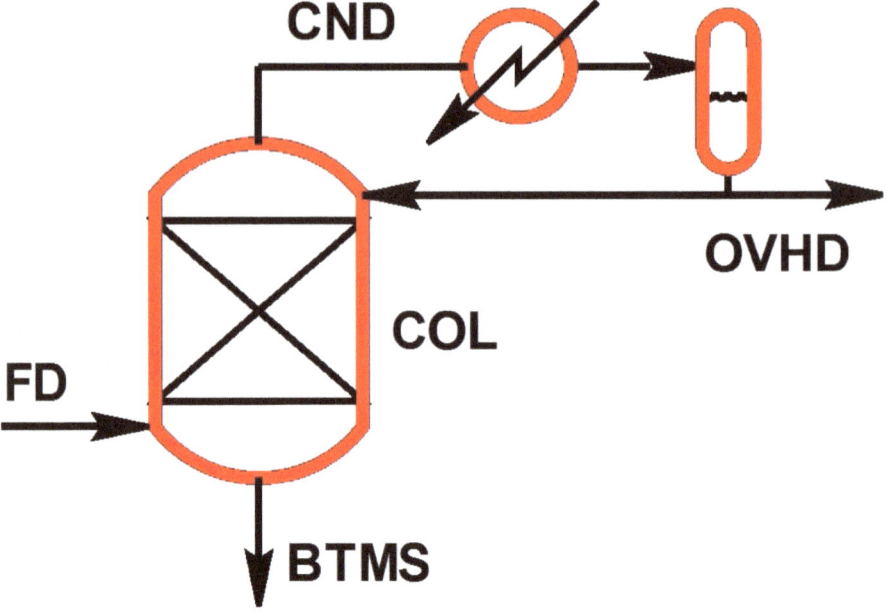

Fig. 2.6

If the feed contains sufficient vapor, and, for certain purity requirements, a separation can be accomplished in a rectifier. This is an example where something was not necessary but it was there because it was necessary in some other situation so the designer followed the convention. Likewise, if the feed contains sufficient liquid, one can eliminate the condenser. This is called a stripper.

What if we remove the condenser and the heating utility from the reboiler?

Fig. 2.7 shows a nitrogen (N_2) stripper. It removes N_2 from liquid natural gas (LNG). A stripper is a distillation column without a condenser. Instead, it has a liquid feed at the top. Since LNG feed is at a cryogenic temperature, it is difficult to efficiently provide an external reboiler utility (such as water or steam used in a conventional column). The utility should be only slightly warmer than the liquid at the bottom of the column. It turns out one can subcool LNG feed to the top of the column against the liquid withdrawn from the bottom. Liquid will boil and provide vapor traffic for the column.

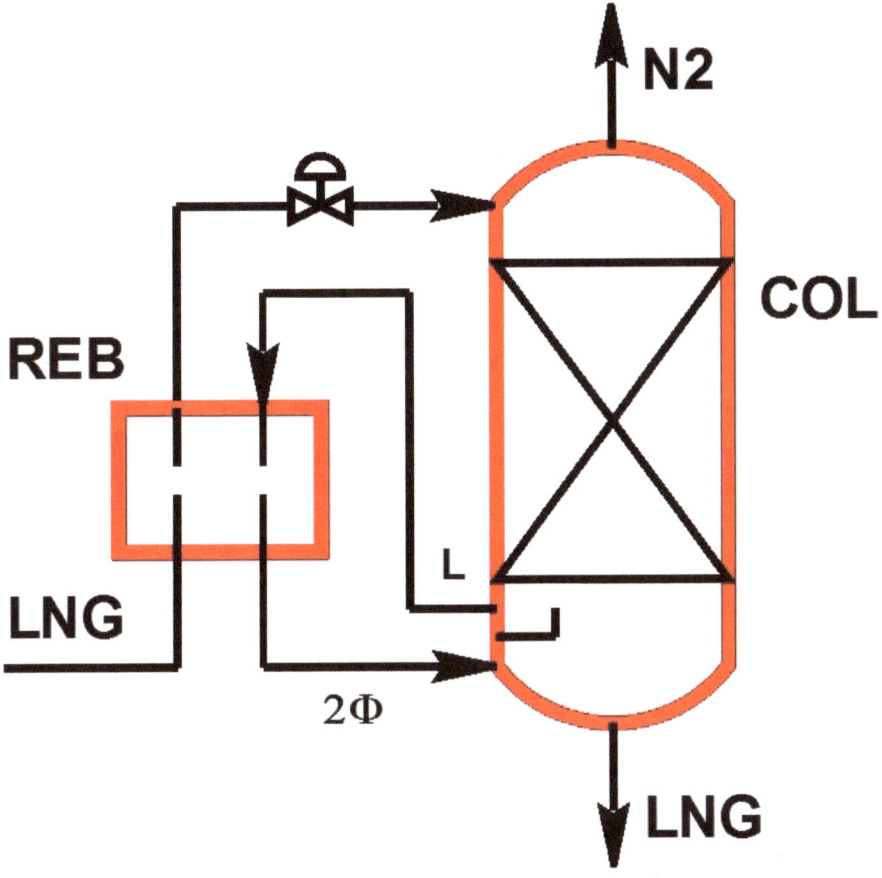

Fig. 2.7

Let us look at a natural gas liquefier (Fig. 2.8). Natural gas (NG) is cooled by propane (C_3) in a series of heat exchangers. It is then introduced to a scrub distillation column (SC) to remove heavy components, so-called natural gas liquid (NGL). It is then partially liquefied in the middle bundle (MB) of the main cryogenic heat exchanger (MCHE) by vaporizing mixed refrigerant (MR). The resulting two-phase stream is separated in a phase separator (PS). The liquid portion is used as the reflux for the column. The vapor portion is totally liquefied in the middle bundle (MB) and subcooled in the cold bundle (CB). It is stored in a tank as liquid natural gas (LNG) product.

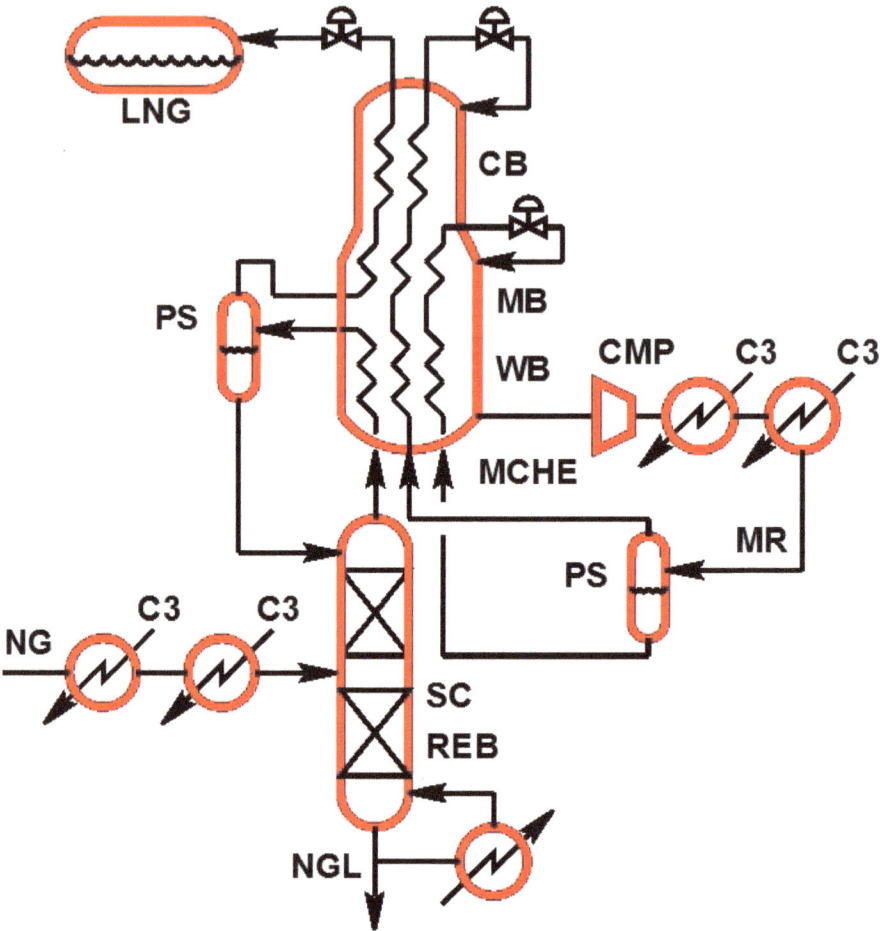

Fig. 2.8

Let us do some radical Trimming/Subtraction (Fig. 2.9). The reboiler is now eliminated. It turns out that, under certain conditions, it can be replaced by a portion of the feed stream at a warmer temperature (in this case the stream is not cooled). It sometimes works better than a rectifier. I call this configuration the "poor man's reboiler." The middle bundle and the phase separator, previously used to generate reflux (liquid) for the column are gone. The new source of reflux is a portion of LNG product. This is the subject of US 2008/0016910 patent application on which I am a coinventor.

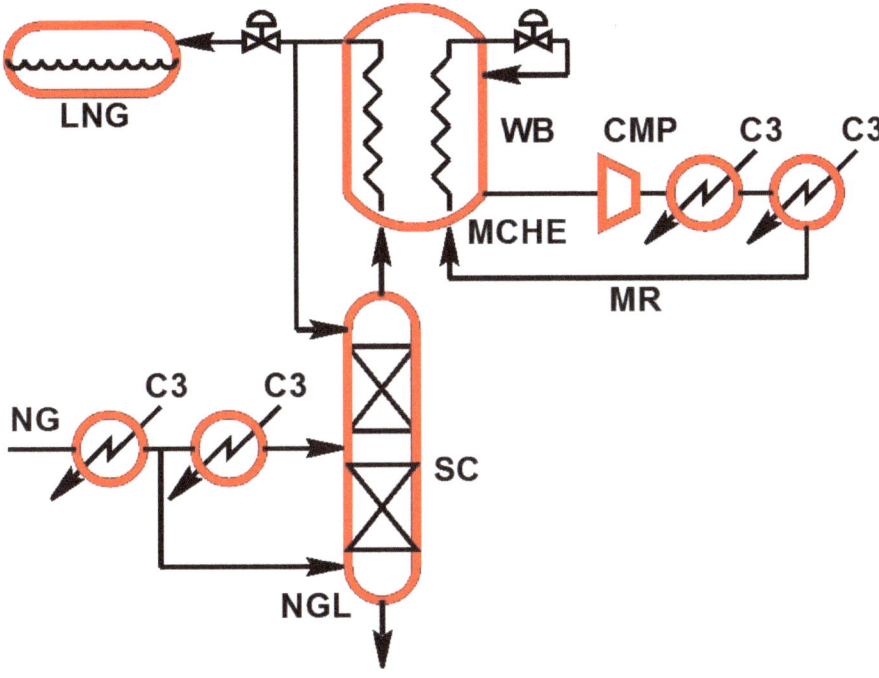

Fig. 2.9

The cold bundle is eliminated. It is normally used to assure that the LNG product can be stored cold at a relatively low pressure. In this case, it is stored at a higher pressure. This requires modification to an existing component of the system—the walls of the storage tank are made thicker.

Let us consider a cryogenic air separation unit (ASU) producing gaseous oxygen product (GOX), shown in Fig. 2.10. Air is compressed in the main air compressor (MAC), cooled in the aftercooler (AC), cleaned up in mole sieves (MS, not shown in Fig. 2.10 for simplicity, an example of conditional Subtraction; shown in Fig. 2.14), cooled in the main heat exchanger (HX), and fed to the high-pressure column (HP).

A portion of air is expanded in the expander (EXP) with an attached generator (GEN) to produce refrigeration and recover some power. GOX is withdrawn from the low-pressure column (LPC), warmed up in HX to recover the refrigeration, and compressed to the desired pressure in a product compressor (CMP).

Waste nitrogen (WST) is also withdrawn from the LPC, warmed up in the HX, and used to regenerate the mole sieves (there are two mole sieve beds—one on stream (used), the other one being regenerated). HP and low-pressure (LP) columns are in thermal communication through the condenser-reboiler (REB) driven by condensing nitrogen, some of which is recovered as gaseous nitrogen product (GAN).

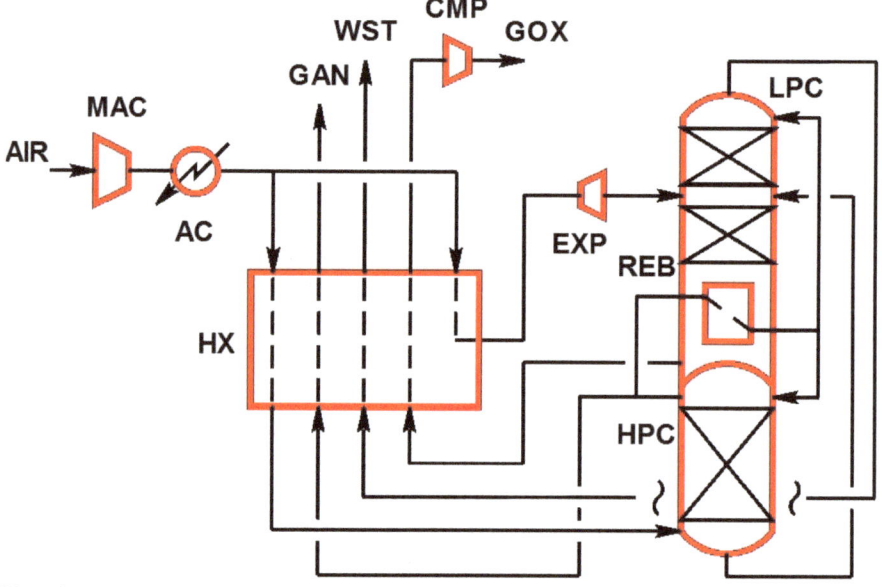

Fig. 2.10

Let us try to remove each component in turn, starting from the MAC. Let us assume there is a large gas turbine (GT) operating (or planned for) in the vicinity of the ASU (Fig. 2.11). A high-pressure slip stream from the gas turbine now becomes the source of air for the ASU. We always want to identify available resources.

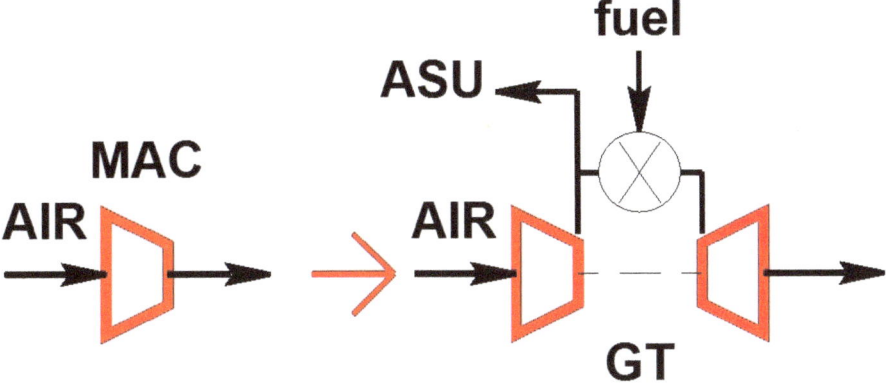

Fig. 2.11

Can we remove intercoolers (not shown) or an aftercooler (AC)? Yes, if efficiency is not an issue, one can consider so-called adiabatic MAC (as opposed to more reversible isothermal compression).
   Can we remove the phase separator?

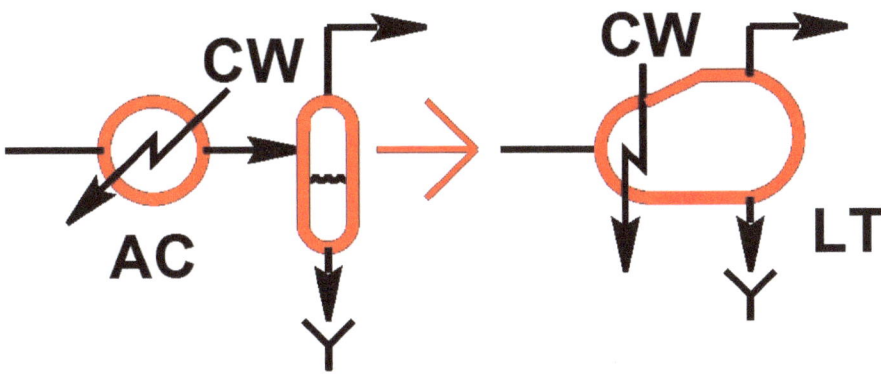

Fig. 2.12

The aftercooler is typically a shell-and-tube heat exchanger. One can use part of the shell for liquid (condensed water) disengagement. A liquid trap (LT) removes water (Fig. 2.12)
   Can we get rid of the tubes and the phase separator? Now we have a spray tower (ST). Cooling water is directly contacted with air (Fig. 2.13). Paradoxically adding water removes water!

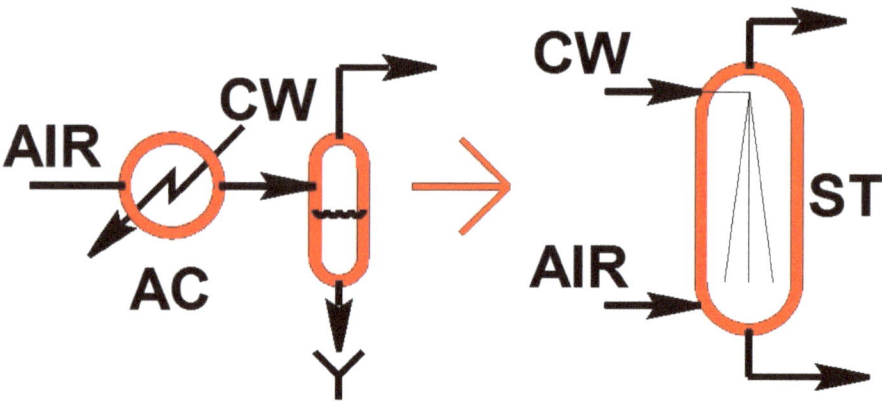

Fig. 2.13

Can we remove the low-pressure column? Yes. We have a gaseous nitrogen (GAN) generator (Fig. 2.14). Now the waste contains about 35 % O_2.

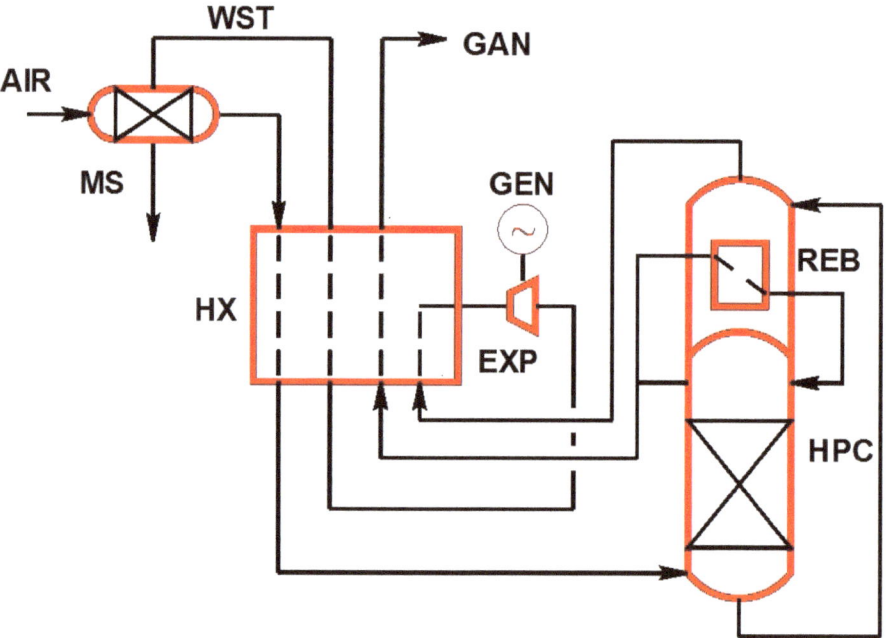

Fig. 2.14

Can we remove the molecular sieves (MS)? Yes. We can freeze CO_2 and H_2O in the main exchanger and then blow it clean (reversing heat exchangers). Can we eliminate the main exchanger HX? Yes. It is the subject of U.S. 6,487,877 (Nitrogen Generation Process); I am the coinventor.

Can we remove the high-pressure column? I do not know. Do you?

Subtraction often removes elements that are no longer useful, but they are in place for historical reasons. It is always good to look into the history of the object, device, or process in question. For example, traffic on the left side of the road in Britain comes from the times when two knights wielding swords in their right hands would pass each other ready for action. Today most people do not carry swords.

Here is an example of Subtraction applied to an entirely different discipline: writing. Consider this sentence: "A cryogenic air separation process is a process that separates air to produce oxygen and nitrogen molecules." Let us trim it a bit: "Cryogenic air separation produces oxygen and nitrogen." We got rid of many unnecessary words without changing the meaning of the sentence or reducing the amount of information.

In one of my previous books, I wrote: "When I was in college I had a Chinese roommate who was a graduate business student." Let us apply Subtraction. The final version would be: "In college I had a Chinese roommate, a graduate business student."

UNIFICATION (CONSOLIDATION, COMBINE)

In SCAMPER its term Combination means merge, hybridize, integrate; mix components, processes, systems.

SIT calls this fairly obvious principle by the term Unification. TRIZ calls it Consolidation (Principle 5): two or more tasks/objects/elements are combined to perform multiple duties. Objects performing contiguous operations are consolidated in space. Contiguous operations are consolidated in time.

It often naturally follows Subtraction. For example, the snakes mentioned in the previous section had replaced its legs with the winding body movement.

Unification was used to arrive at solutions shown in Figures 2.7 (column feed doubles as the reboiler utility), 2.12 and 2.13 (see section on Subtraction).

An obvious trivial example of Unification is a shampoo with conditioner.

I use this principle in a least in two ways: (a) to merge parts of an existing system or process to simplify it, as explained above; (b) to merge/hybridize different objects or processes to come up with a new invention/application.

Fig. 3.1

Fig. 3.1 shows a mule, a hybrid of a horse and a donkey. Mules are big and strong as horses and smart, unafraid of heights, and not picky about food like donkeys.

Here is an example of a new technical invention. Can we combine an airplane and a balloon?

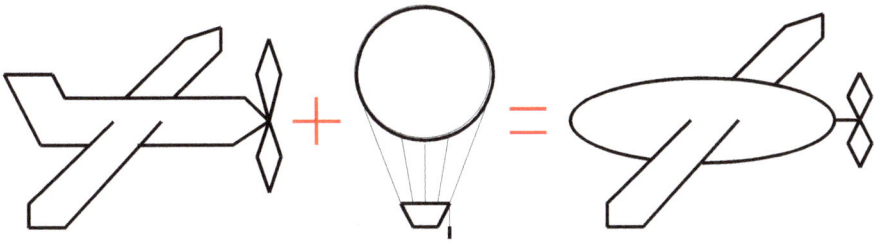

Fig. 3.2

Yes, we have a relatively new invention, the Dynalifter airship (Fig. 3.2). Unlike dirigibles and blimps, Dynalifters are heavier than air. They need little space to land and to take off, and they can carry heavy loads.

Fig. 3.3 shows an example from Engineers Without Borders project in Rwanda showing an improvement to a certain application (as opposed to a new invention) obtained by merging different technologies.

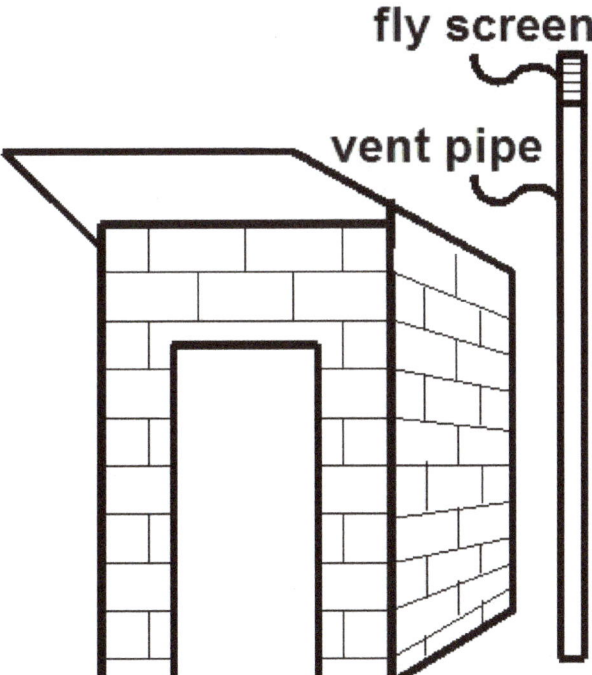

fly screen

vent pipe

Fig. 3.3

The ventilated improved pit (VIP) latrine is constructed according to original guidelines but using the local adobe bricks as opposed to conventional wood or cardboard as building material. The design (a) takes advantage of the local expertise with adobe in Rwanda and (b) makes the construction more robust in the seismic zone.

Can we combine a bicycle and a parachute?

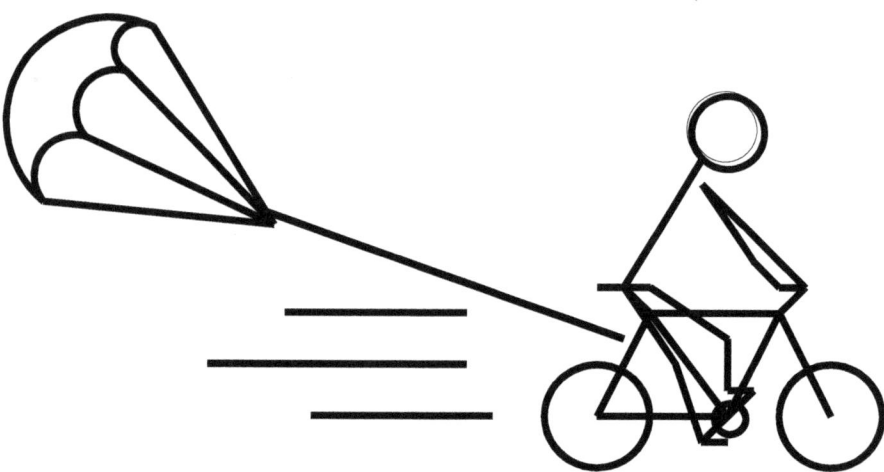

Fig. 3.4

Fig. 3.4 shows a new training method before Tour de France. Can we combine them in a different way? Maybe (Fig. 3.5).

Fig. 3.5

SUBSTITUTION

Substitution is a SCAMPER principle (the *S* in the acronym) that has no direct TRIZ equivalent. Like Unification, it naturally follows Subtraction.

I use it in two ways: (a) to perform the same function as the original device, system, or process, but with an old element replaced by a new one; and (b) to come up with an entirely different application or invention.

Bicycle and motorcycle, shown on Fig. 4.4, perform the same function but with the pedals replaced by an engine.

Let us look at a parachute. What if we substitute a person with an airplane (different application)? High-profile accidents of small airplanes, such as John Denver's and JFK Jr.'s, convinced some companies to install parachutes that would slow down the airplane's fall in case of a total loss of control (Fig. 4.1). This is a relatively new implementation.

Fig. 4.1

Fig. 4.2 shows a dirigible. What if we leave the device alone and substitute the medium?

Fig. 4.2

If we substitute air with water we have a bathyscaphe, a new invention (Fig. 4.3).

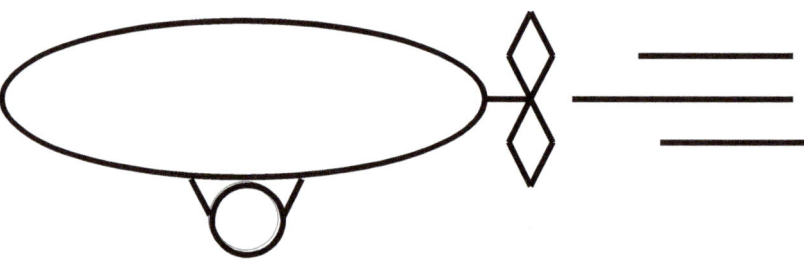

Fig. 4.3

In section on Subtraction I mentioned skateboard and surfboard. I can add to this list skyboard. We have substituted ground with snow then with air.

Fig. 4.4

40 INVENTIVE PRINCIPLES, SIT, AND SCAMPER

In previous sections I talked about some specific principles. If one wants to learn to swim I think it is better to get in the water first and try some stokes instead of learning the theory on land. It is also more fun. Now I am going to explain some theory.

One of the main TRIZ tools is the 40 Inventive Principles. They are:

1. Segmentation
2. Extraction
3. Local Quality
4. Asymmetry
5. Consolidation
6. Universality
7. Nesting
8. Counterbalance
9. Prior Counteraction
10. Prior Action
11. Cushion in Advance
12. Equipotentiality
13. Do It in Reverse
14. Spheroidality
15. Dynamicity
16. Partial or Excessive Action
17. Transition into a New Dimension
18. Mechanical Vibration
19. Periodic Action
20. Continuity of Useful Action
21. Rushing Through
22. Convert Harm into Benefit
23. Feedback
24. Mediator
25. Self Service
26. Copying
27. Disposable
28. Replacement of Mechanical System
29. Pneumatic or Hydraulic Construction
30. Flexible Membranes or Thin Films

31. Porous Material
32. Changing the Color
33. Homogeneity
34. Rejecting and Regenerating Parts
35. Transformation of Properties
36. Phase Transition
37. Thermal Expansion
38. Accelerated Oxidation
39. Inert Environment
40. Composite Materials

I am not going to discuss all of them detail. I refer the reader to the suggested reading section herein entitled Literature. We have already discussed Principle 2: Extraction, and Principle 5: Consolidation. I am going to discuss some of the other principles in the following sections.

As I mentioned before, the principles were created starting in the 1940s in the former Soviet Union. If these principles had been instead created in the United States some ten years ago, then maybe, instead of Pneumatic Construction or Flexible Membranes, we would have Nanotechnology. Some other principles are more universal and, not surprisingly, correspond to similar principles in other idea-generation techniques/creativity tools.

Some TRIZ gurus say one can apply any of those forty principles to any type of problem. For example, one can apply Pneumatic Construction or Flexible Membranes to business problems. I personally find such statements silly. Like I said before, if one puts together a bunch of smart people in one room, one will get some solutions.

Let us apply TRIZ to TRIZ. Let us unify or consolidate (Principle 5) some principles. Principle 35 (Transformation of Properties) and Principle 36 (Phase Transition) can be merged into one: Transformation. Principle 1 (Segmentation) can be applied to arrive at Principle 13 (Do It in Reverse) or Principle 10 (Prior Action). See subsequent sections herein.

SIT was created 1990s in an attempt to pick the most useful TRIZ principles, with some modifications, reducing the number of principles to five: Subtraction, Unification,

Multiplication, Division, and Attribute Dependency. As mentioned before, Subtraction is similar to TRIZ's Extraction (Principle 2). Division is related to Segmentation (TRIZ Principle 1) and Unification to Consolidation (TRIZ Principle 5). There is no direct TRIZ (or SCAMPER) equivalent to Multiplication. The last principle, Attribute Dependency, is broader.

In the 1950s Alex Osborn came up with a list of ten creativity steps. They include:

> A) Put to different use! Can this solution be applied to a different discipline or application?
> B) Adapt! Can a solution from a different domain be used?
> C) Change, modify! Can one change shape, color, etc.?
> D) Enlarge!
> E) Reduce!
> F) Replace, substitute!
> G) Rearrange!
> H) Invert, reverse!
> I) Combine!
> J) Transform!

Then in the 1960s Bob Eberle simplified the Osborn list to produce SCAMPER:

> S) Substitute
> C) Combine
> A) Adapt
> M) Modify
> P) Put to Other Use
> E) Eliminate
> R) Reverse, Rearrange

As mentioned before, Osborn's list and SCAMPER were created independently of TRIZ and SIT. As one would expect, there is significant overlap between these techniques.

I am not sure I agree with Altshuller's notion that anyone can become an inventor. Some people are naturally very creative and use TRIZ–like or SCAMPER–like techniques implicitly, without thinking about it. Other creative thinkers would reject the notion that it can be taught. Some people are just not creative and never will be. Other people are not normally creative but can benefit from a structured approach. I think creativity tools and idea-generating techniques can make some but not all creative people more creative and can make some but not all noncreative people somewhat creative. It is possible that the majority can benefit from such tools and techniques.

MULTIPLICATION

This is a powerful SIT principle. There is no direct TRIZ or SCAMPER equivalent. TRIZ principles of Heterogeneity and Composite Materials are somewhat related.

Multiplication suggests making a copy of the object modified in some way (this is different from TRIZ's Copying principle). Typically a modified copy performs a slightly different or completely different function.

Simple examples include a toothbrush with different types of bristles, a razor with different types of blades (one to lift the hair; the other one to cut it), and bifocals (invented by Benjamin Franklin).

An aviation example is shown in Fig. 5.1:

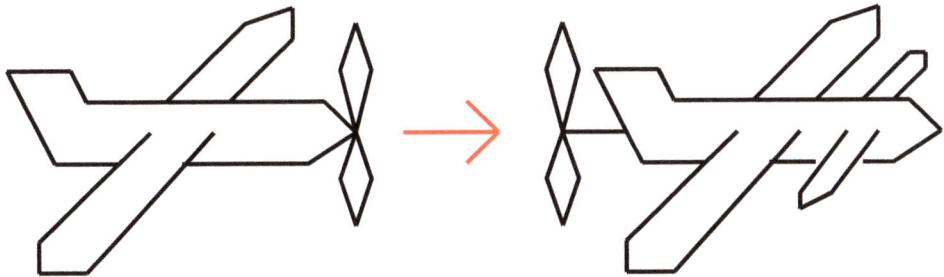

Fig. 5.1

A wing of a conventional airplane is copied and modified—reduced in size. This improves stability of a new type of airplane invented by Burt Rutan.

Let us apply Multiplication to a parachute (Fig. 5.2).

Fig. 5.2

The little "pilot" chute helps open the bigger one.

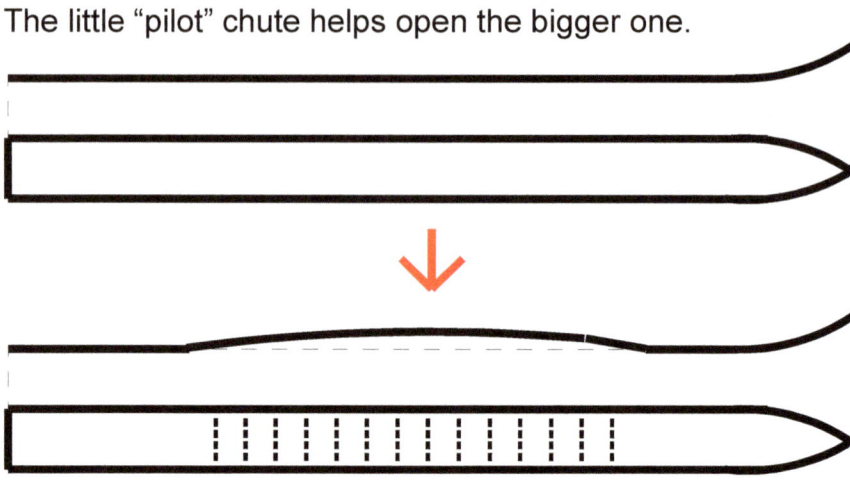

Fig. 5.3

I have a pair of backcountry skis (Fig. 5.3). Conventional skis are homogenous. My skis are convex in the middle. At the same time, the bent section's surface is like a rasp. When I move one ski forward only the smooth parts of the ski touch the snow. When I rest my weight on the other ski, the portion bent upward straightens and provides more friction. Part of the ski was multiplied and modified.

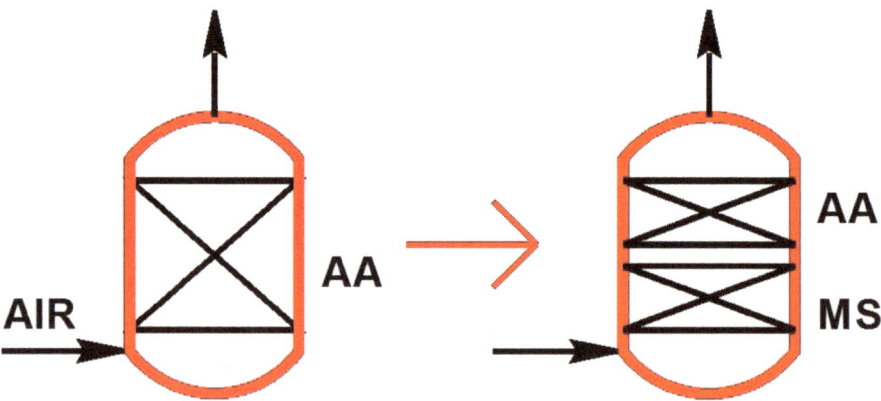

Fig. 5.4

Let us consider a TSA (temperature swing adsorption) unit to remove water and CO_2 from air (Fig. 5.4).

The adsorbent used is activated alumina (AA). Let us multiply the bed and modify one of the beds to contain molecular sieves (MS). It turns out the new unit removes CO_2 and water more efficiently.

Let us consider a distillation column once again (Fig. 2.5). Let us multiply the reboiler and add an intermediate reboiler (IREB, Fig. 5.5). Now some of the steam used to drive the column can be supplied cheaply at a colder temperature, thus reducing the hot steam requirement. Likewise, we can add an intermediate condenser to reduce the cooling utility (such as chilled water) requirement.

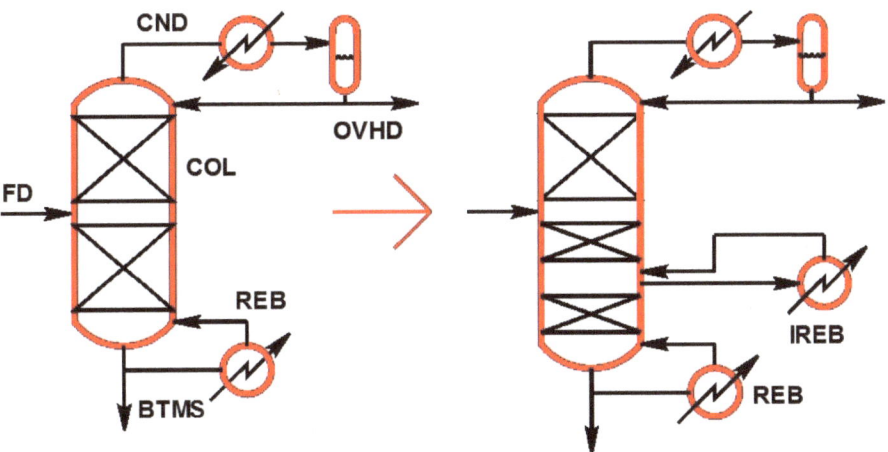

Fig. 5.5

Let us apply the intermediate reboiler to a nitrogen stripper (see Fig. 2.8). This means multiplying and modifying the existing reboiler (see Fig. 5.6). This configuration is patented.

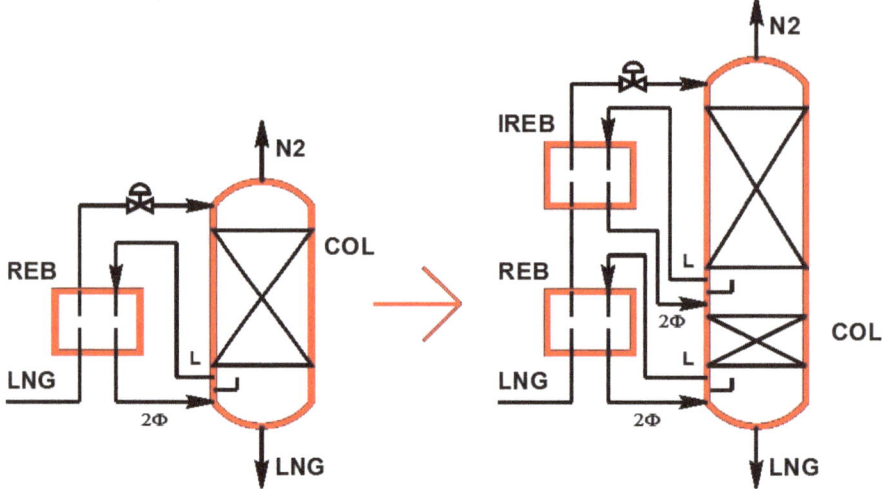

Fig. 5.6

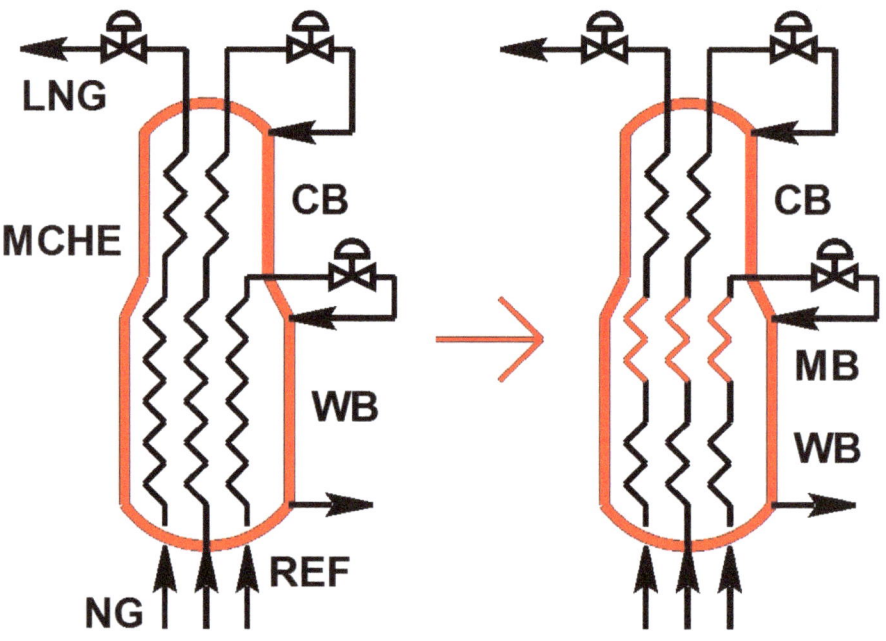

Fig. 5.7

35

Here is an example from natural gas liquefaction technology (Fig. 5.7). The wound-coil main cryogenic heat exchanger (MCHE) on the left has two tube bundles: the warm bundle (WB) to liquefy the natural gas and the cold bundle on top to subcool the product. The warm bundle has to operate over the wide range of conditions: from chilled gas to cryogenic liquid.

Let us multiply the warm bundle (Fig. 5.7). The exchanger on the right now has an additional bundle: middle bundle (MB). WB and MB have different geometry: different mandrel diameter, number of tubes, winding angle, etc. The modification yields better heat transfer performance and higher LNG production from the exchanger of a given size (at the cost of more complexity). The invention is the subject of European Patent EP 1 367 350.

The idea was not a result of using any idea-generating techniques; it was a result of intuitive thinking. It could have possibly been generated much earlier, if a technique like SIT's Multiplication had been applied.

Please notice that Multiplication seems to be the opposite of Subtraction. Sometimes it is better to remove things, sometimes to add things, even in the same type of device or operation.

Here is an example of Multiplication applied to another engineering discipline.

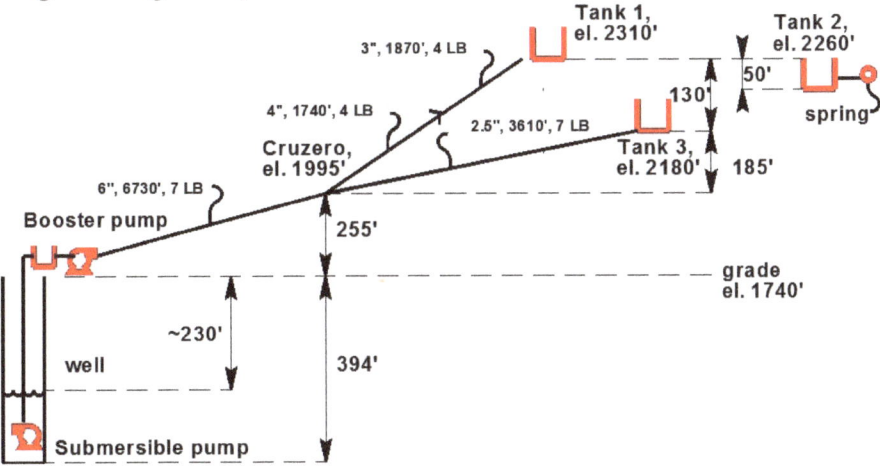

Fig. 5.8

Fig. 5.8 shows an existing clean water system in a village of Las Delicias in El Salvador. The Engineers Without Borders project in Las Delicias, El Salvador, is a retrofit. The existing system consists of a submersible pump, a booster pump, two storage tanks at relatively high elevation (the third tank is supplied by a spring), and gravity distribution piping which supplies water to over three thousand people. The village, located on the flank of a volcano, cannot afford the current energy bill.

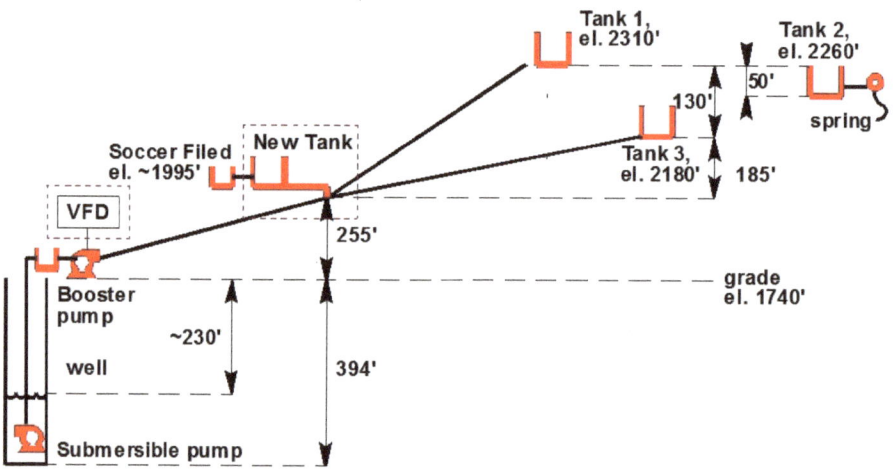

Fig. 5.9

Our team concluded we could save the energy required to pump water or, conversely, supply about 30 % more water for the same price, by adding another tank at an intermediate elevation (Fig. 5.9). The tank will supply about 50 % of the houses located at lower elevations.

In this case, Multiplication was applied to the storage system. The modified feature is a new tank's lower elevation.

What if we multiply the pump? Fig. 5.10 shows a solution where another pump at a higher elevation is added to the new tank. This solution was not used in the final design.

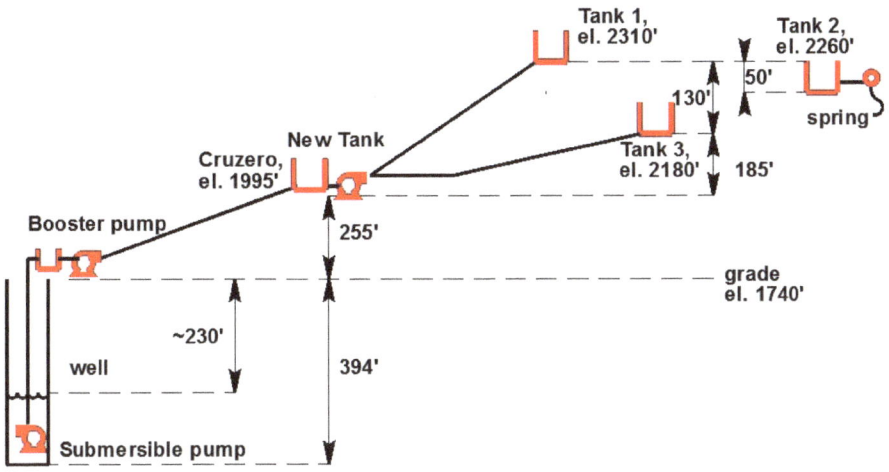

Fig. 5.10

As a side note, according to WHO about one in nine people worldwide lack access to safe water and 1.8 million people die every year from diarrheal diseases resulting from inadequate drinking water; 90% are children under five, mostly in developing countries. According to Water.org, the water crisis is the number one global risk based on impact to society and number eight based on likelihood of occurring within 10 years.

Multiplication is also related to the TRIZ Principle 3: Local Quality, including making homogenous objects heterogeneous.

SEGMENTATION (DIVISION)

SIT's Division principle calls for dividing a system into parts and, optionally, rearranging the parts. For example ABC is divided into A-B-C. The existing options are A-B-C, B-A-C, A-C-B, C-A-B, C-B-A, and B-C-A (6! Possibilities). The SIT principle talks mostly about physical systems. Here we want to extend it to actions and processes.

TRIZ's related Segmentation principle (Principle 1) does not normally call for rearrangement. It talks about dividing an object into parts.

The Division principle is related to SCAMPER's *R* (Reverse/Rearrange): change sequence, order, make moving part stationary. The Reverse part will be covered separately.

Fig. 6.1 shows a conventional airplane being modified. Let us name the parts attached to the fuselage in sequence: tail-wings-propeller. The modified airplane design has a different sequence: propeller-tail-wings.

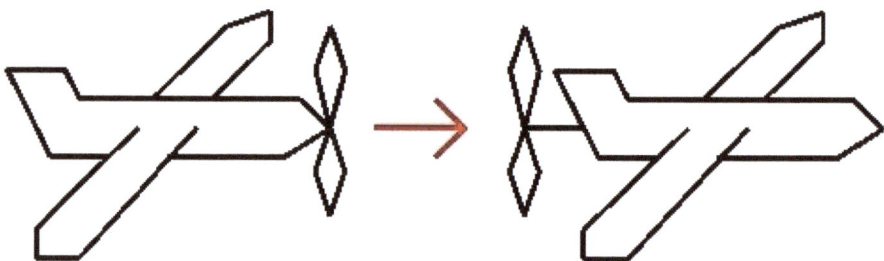

Fig. 6.1

Division can be used to arrive at TRIZ principles of Prior Counteraction, Prior Action, and Cushion in Advance. Those are TRIZ principles 9, 10, and 11. To me, they are essentially the same. They advocate taking action in advance to (a) improve the outcome and (b) to prevent a negative outcome. Let us look at the process that involves steps A-B-C-D. Can we do step C first to make steps A and B easier?

Here is an example. I am making scrambled eggs. The steps are: heat the frying pan (A), pour eggs (B), mix them in the frying pan (C), turn off the heat (D), add salt (E) and serve.

If I add salt later, I risk not distributing salt evenly. So I heat the frying pan (A), add salt while mixing raw eggs (E-C), turn off the heat (D: take advantage of thermal inertia of the pan and save power), pour them in the frying pan (B), then serve. This way I achieve a positive outcome (seasoned eggs), and, at the same time, I prevent negative outcomes.

     Prior Counteraction and Prior Action may diverge from Division if an additional step is introduced prior to other steps. One way to look at it in more general terms is that every process should start with a wildcard step called Prepare which may or may not be utilized.

     Fig. 6.2 shows segmentation followed by rearrangement applied to the system shown in Fig. 2.9. Natural gas feed is precooled by at least two stages of propane (C_3)—in this case Stage 1 at a warmer temperature and Stage 2 at a colder temperature. The sequence is Stage 1-Stage 2-Scrub Column-MCHE (liquefaction). What if we rearranged the sequence? We now have Stage 1-Scrub Column-Stage 2-MCHE. Stage 2 of the propane segmentation now provides reflux to the column.

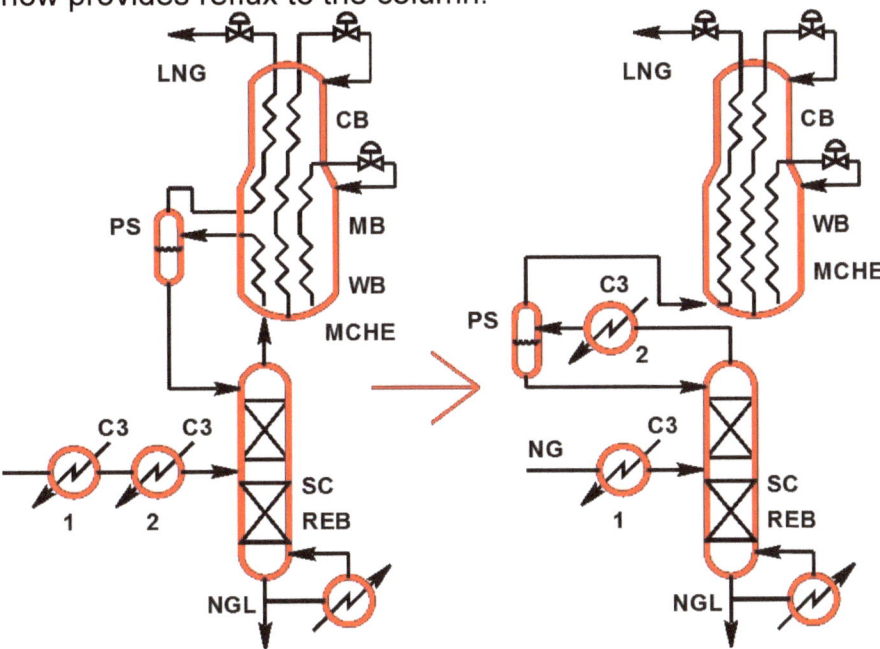

Fig. 6.2

REVERSE ACTION

When I was maybe ten, I attached a small lightbulb to an electric motor that I had removed from a toy car. Then I started to manually turn the motor. I intuitively felt that, since supplying electricity to a motor would make it turn, I could generate electricity by my opposite action. My experiment did not work too well, but, it turns out, there are electric motors that can become generators.

When Polish-born astronomer Copernicus was in a boat he realized it seemed the boat was stationary and the river bank was in motion so he stopped the Sun and moved the Earth!

The Reverse Action principle is present in TRIZ (Principle 13 - Do It in Reverse: implement the reverse or opposite action to solve a problem) and SCAMPER (as part of Reverse/Rearrange: change sequence, order, make a moving part stationary). It may mean lowering object A instead of raising object B, making object A smaller instead of making B bigger, making A hot instead of making B cold.

The way I apply it, the Reverse Action principle can be used in at least two ways: (a) to solve a problem by Reverse Action and (b) to come up with a new invention by performing the action opposite to some known action.

Solving a particular problem by an opposite action or by make a moving part stationary and vice versa seems fairly obvious.

Here is a wind power example (Fig. 7.1).

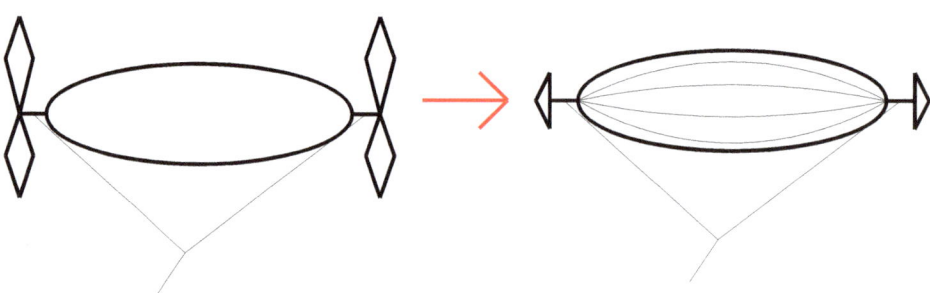

Fig. 7.1

On the left, Fig. 7.1 shows a system from page 102 of the September 2006 special issue of *Scientific American*. It shows a balloon with wind turbines and generators attached. Let us eliminate wind turbines and replace them with something already existing which can be put to another purpose. We have a more revolutionary approach: floating wind generators. The previously stationary helium-filled balloon catches the wind in fabric scoops and turns generators at each end.

The book entitled *40 Principles: TRIZ Keys to Technical Innovation* by Genrich Altshuller talks about a fish farm with compressed air injected into the pond. Fig. 7.2 shows another solution.

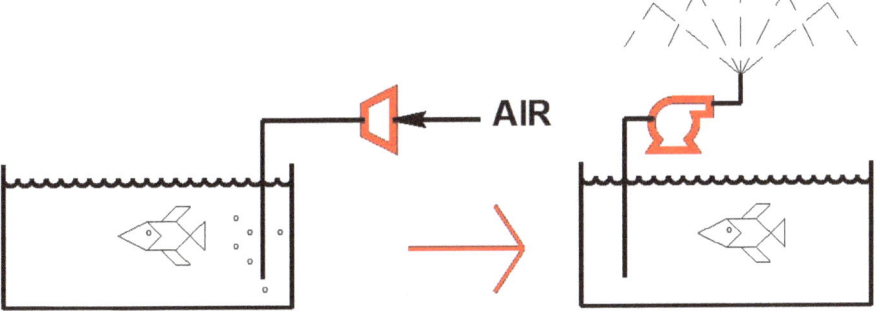

Fig. 7.2

Instead, one can spray water into the air using a pump. Evaporative cooling improves solubility of oxygen in water.

I extended the principle to come up with a new invention or a new application.

For example, one can use a wristwatch with hands on a sunny day to determine the southern direction. Conversely, one can use a compass as a portable sundial. Electrolysis has been known for a long time. Fuel cells, although invented almost two hundred years ago, became widely used relatively recently. They perform the exact opposite action.

Parasailing is a popular beach pastime. What if, instead of dragging a parachute canopy behind a boat, we use a canopy to pull a ship? We have a SkySail, a new invention (Fig. 7.3). Installing SkySails on a tanker saves up to 20 % of its fuel. Anybody could come up with this invention just by playing with the building blocks. Then it takes somebody with insight and experience to find an application.

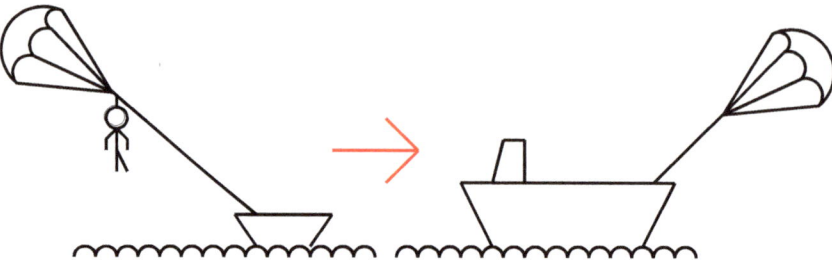

Fig. 7.3

Everyone knows that if one mixes two streams, a cold stream and a hot stream, for example, in a water faucet, one will get a stream at an intermediate temperature. Can one "unmix" a single stream into two streams at two different temperatures? It turns out such a device exists. It is called a Hilsch-Ranque Vortex Tube (Fig. 7.4).

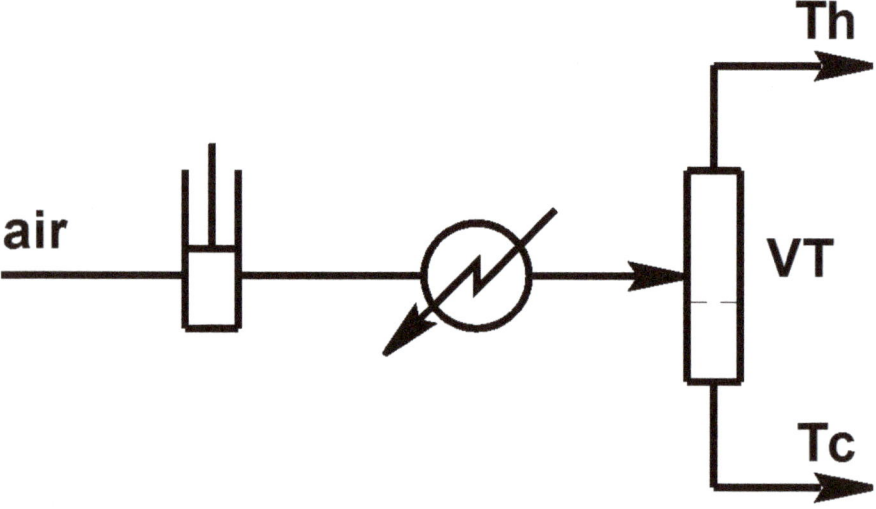

Fig. 7.4

A vortex tube (VT) splits a high-pressure gas stream into two low-pressure streams, hot and cold, on the principle of kinetic energy conversion. Air is compressed, optionally cooled to close-to-ambient temperature, and expanded to atmospheric pressure through a vortex tube. The air is introduced tangentially, producing a swirl. There is an orifice across the tube. Only molecules with low kinetic energy cross the orifice. This produces hot and cold streams at temperatures Th and Tc.

The Reverse Action principle can be used to solve problems. Fig. 7.5 shows a method of calculating the distance to the horizon. It can be easily derived as:

AB ≈ sqrt (2 R h)

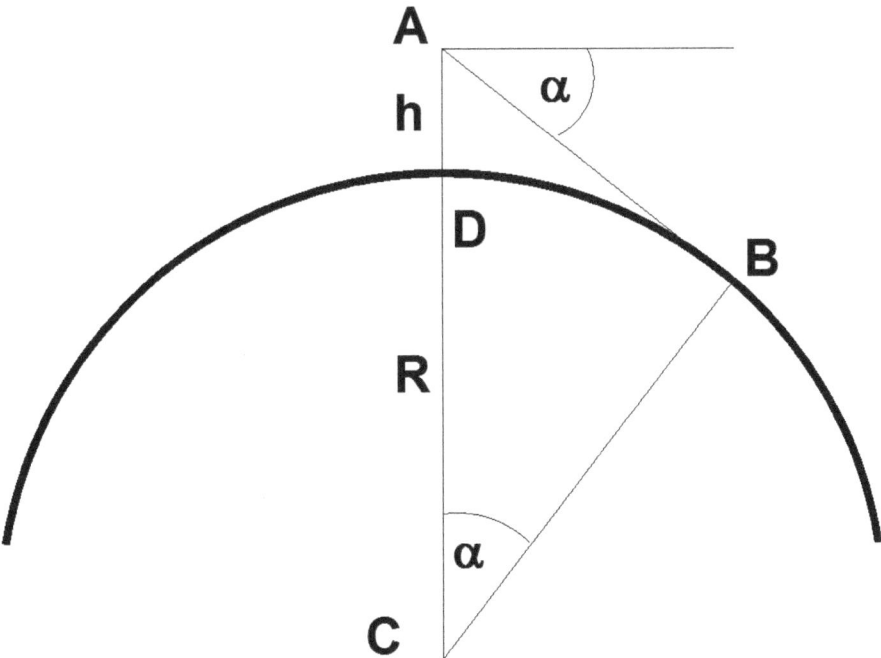

Fig. 7.5

Let us do something opposite. We want to know the distance from the Rock of Gibraltar to the Moroccan coast. We are climbing the rock until we see the coast on the horizon. We can measure the altitude (e.g., using a barometer) and get the distance. Or, if we know the distance (about 33 nautical miles), we can calculate the radius, and, by extension, the circumference of the Earth.

Can we "undistill" a fluid?

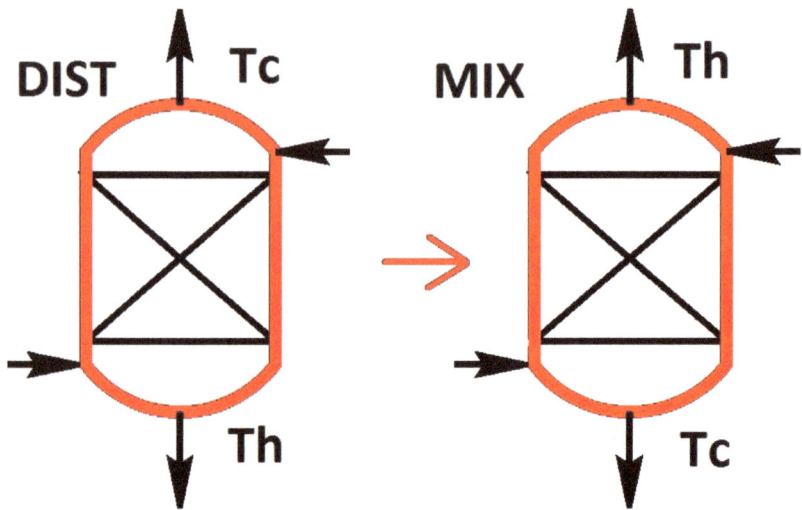

Fig. 7.6

Fig. 7.6 shows such a device, the mixing column (MIX). Mixing fluids of different temperature, pressure, or composition is inherently inefficient. The distillation column separates components. The top of distillation column (DIST) is the coldest (Tc); the bottom is the hottest (Th). The mixing column reversibly mixes fluids. The top of mixing column is the hottest; the bottom is the coldest. The idea is not very intuitive but one could come up with it by applying Reverse Action.

An important application of the Reverse Action principle is finding an application for a device or an idea instead of finding a device or a method to solve a specific problem.

Here is a puzzle supposedly given at job interviews: how to uncork a wine bottle without breaking the bottle or manually removing the cork? Apply Reverse Action. Push the cork into the bottle.

CONVERT HARM INTO BENEFIT

In his memoirs, Polish adventurer Tony Halik wrote about crossing the Americas from Tierra del Fuego to Alaska. The roads were sometimes terrible. He would put his dirty clothes in a drum, add water and soap, and rely on road bumps to do the washing. Steinbeck described a similar technique in *Travels with Charley in Search of America*.

Let us go back to Fig. 2.14, modified as Fig. 8.1.

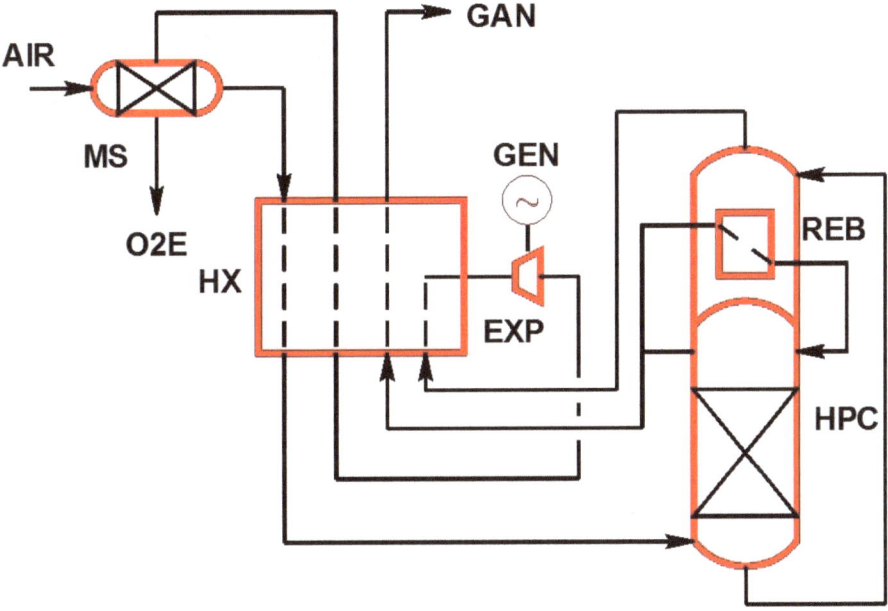

Fig. 8.1

The waste stream contains about 35 % oxygen, compared to 21 % in the atmospheric air. It can be used in oxygen-enhanced combustion (O2E, Fig. 8.1). It does not have to be dry, as combustion produces water vapor anyway.

The Convert Harm to Benefit principle can be extended to Convert Waste to Benefit.

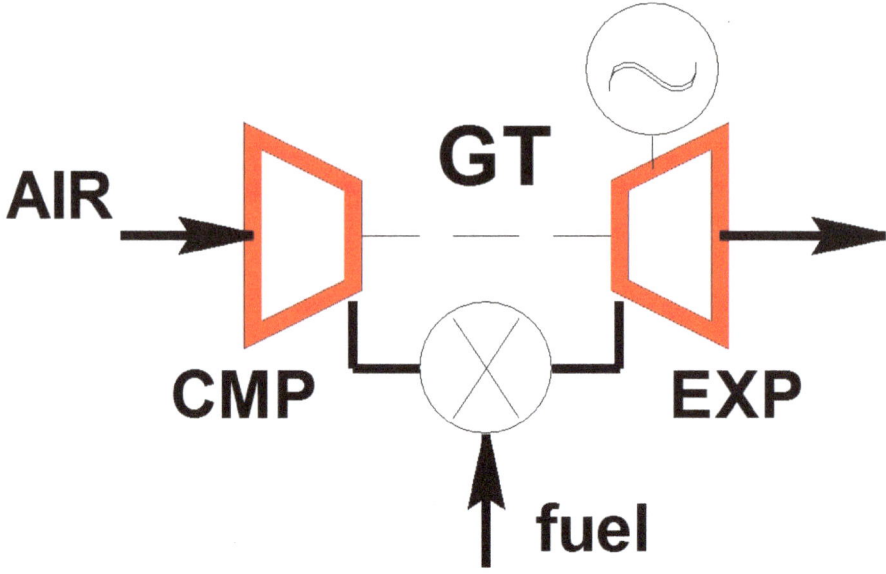

Fig. 8.2

Here is another example of converting waste into benefit. Figure 8.2 shows a gas turbine (GT) generating power. Air is compressed, fuel injected, hot gas resulting from combustion is expanded. The gas leaving the turbine is still quite hot.

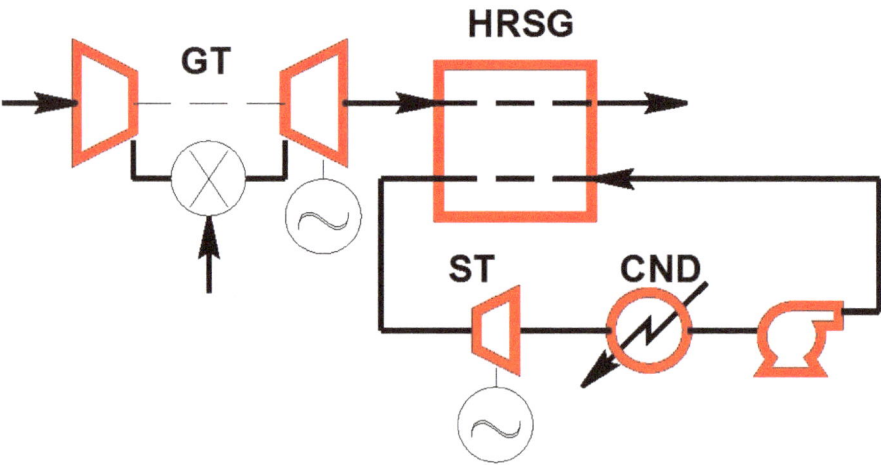

Fig. 8.3

Figure 8.3 shows a combined power-generating cycle. Gas leaving the gas turbine can be used to generate steam in the heat-recovery steam generator (HRSG). The steam expended in the steam turbine (ST) is used to generate additional power. Without the HRSG this heat would be wasted.

In general, in process engineering, if a stream is at above-ambient or below-ambient temperatures, or above-atmospheric or below-atmospheric pressures (vacuum), work or other forms of energy can be extracted.

SELF SERVICE AND THE PATENTING PROCESS

TRIZ principle 25, Self-Service, teaches us to make an object serve itself by performing auxiliary helpful functions.

The distillation column shown in Fig. 2.6 uses relatively warm feed to provide boil-up. This eliminates a heating utility. The double column in Fig. 2.11 eliminates both heating and cooling utility. Somebody observed at the beginning of the twentieth century that oxygen at some low pressure boils at the same temperature as nitrogen at high pressure. This was a breakthrough in air separation. Condensing nitrogen provides reflux for the high-pressure column while boiling oxygen in the low-pressure column to provide boil-up. Thus, the same exchanger acts as both a reboiler and a condenser, performing two functions at the same time.

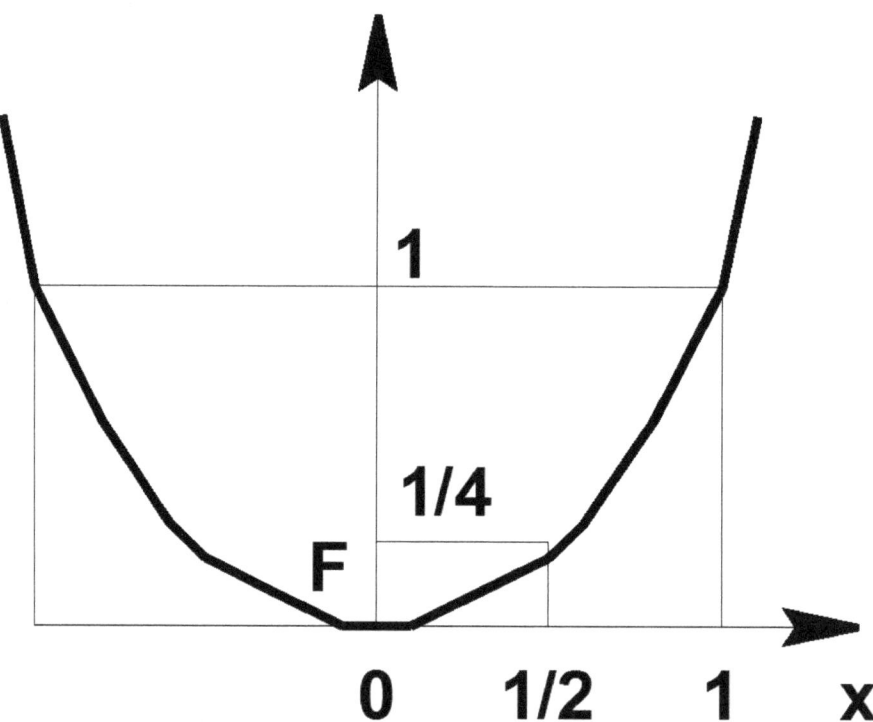

Fig. 9.1

As only one straight line passes through two points, there is a unique parabola passing through three points. If the parabola is given by the equation:

(1) $y = a x^2$,

then the focal length is given by:

(2) $F = \frac{1}{4}a$, and

(3) the focal point has the (x, y) coordinates of (0, F).

Fig. 9.1 shows a parabola of the equation (1) shown above where $a = 1$, $F = \frac{1}{4}$. As two points define a straight line, three points uniquely define a parabola. In addition, a parabola has to be symmetrical. The above parabola is defined by points (0, 0), ($\frac{1}{2}$, $\frac{1}{4}$), and (1, 1). The focal point has coordinates (0, $\frac{1}{4}$).

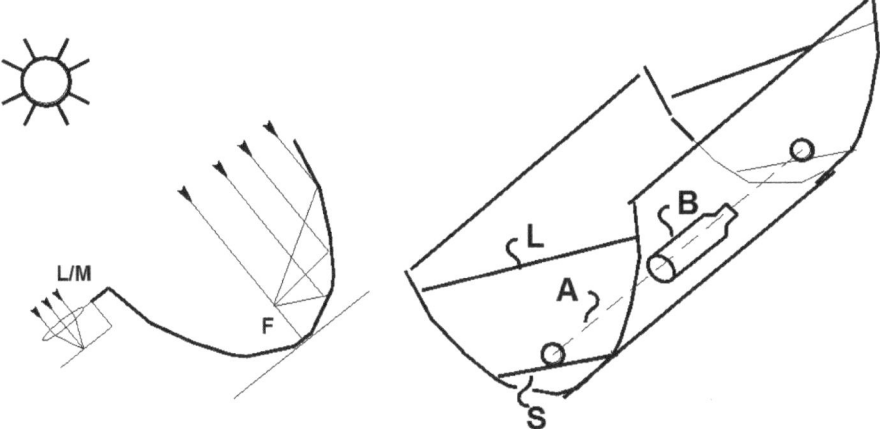

Fig. 9.2

Fig. 9.2 shows a cross section of the parabolic shape of a solar collector with sun rays converging at the focal point F. A small lens or concave mirror (L/M) can be used to orient the trough toward the sun by focusing light on the tangent parallel surface (Multiplication principle again!).

The drawing to the right shows a view of the parabolic trough. Between $x = 0$ and $x = 1$, the length of a parabola given by equation (1), where $a = 1$, is 1.479. Between $x = 0$

and $x = 0.5$, the length is 0.574. That way one can determine the exact location to attach supports L and S.

If one connects points (½, ¼) and (-½, ¼) and points (1, 1) and (-1, 1) at the two ends of the trough, it will assume close-to-parabolic shape at the cross section with the focal axis clearly defined.

Fig. 9.3

One can use four supports connecting those points: two long (L) and two short (S). The line connecting the centers of the short supports is the focal axis (A). If one cuts notches in the supports and the trough, they can be installed without any additional parts. A bottle (B) (dark for pasteurization or even boiling liquid; transparent for improved solar disinfection aka SODIS (Solar disinfection)–type UV treatment of water on a partially cloudy day or with one side painted to enable both uses) is placed along the focal axis.

Long and short supports L and S have a dual role: they provide structural integrity, and they assure close-to-parabolic shape of the trough. In addition, short supports S have a third role: they pinpoint the focal axis A. The device can be used for (a) pasteurization of all liquids, (b) boiling water, (c) cooking, and (d) improved SODIS-like treatment. Multiple bottles can be placed in a long trough.

Fig. 9.3 shows a photo of an improved working prototype with a brown beer bottle in the trough. The trough is made from two layers: cardboard for structural integrity and a thin reflective surface, both cheaply obtained from arts supply stores. The four supports are wooden. Both supports and cardboard/sheet metal are precut log cabin style (notched) for easy assembly without tools. Additional supports can be used to place the bottle exactly along the focal axis, but, because of the short focal length, the bottle can simply be laid inside. The device can be propped to face the sun.

Fig. 9.4 shows a detailed diagram of the system shown in Fig. 5.8. The variable frequency drive (VFD) regulating the speed of the booster pump requires cooling. We have an existing submersible pump and a source of cold water (the well). We can take advantage of this synergy and redirect only about 1 % of the flow to cool the VFD.

The Self-Service principle is important in intellectual property protection as performing auxiliary helpful functions strengthens patent claims. For example, supports L and S in Fig. 9.2 serving multiple functions could make the device more patentable. Instead, I published a disclosure on this subject to make the invention available to anyone.

A classic example of Self-Service in process engineering is the use of an economizer heat exchanger to reduce cooling or heating duty. If a stream has to be cooled for some low-temperature process, it makes sense to precool it against the stream leaving the same operation. Conversely, if the stream is to be heated for some high-temperature process, it makes sense to preheat it against the effluent.

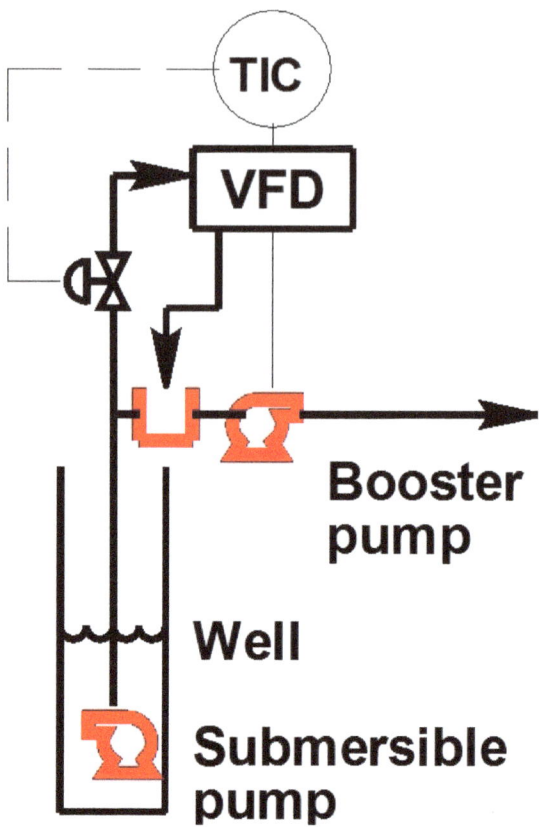

Fig. 9.4

Fig. 9.5 shows the economizer heat exchanger 1 reducing duty of cooler 2 cooling a stream to the low-temperature process 3 by recovering cold of the returning stream.

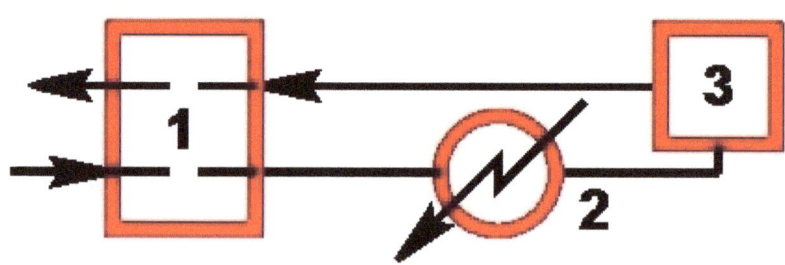

Fig. 9.5

Let us go back to the gas turbine shown in Fig. 8.3, but with modifications as shown in Fig. 9.6 below.

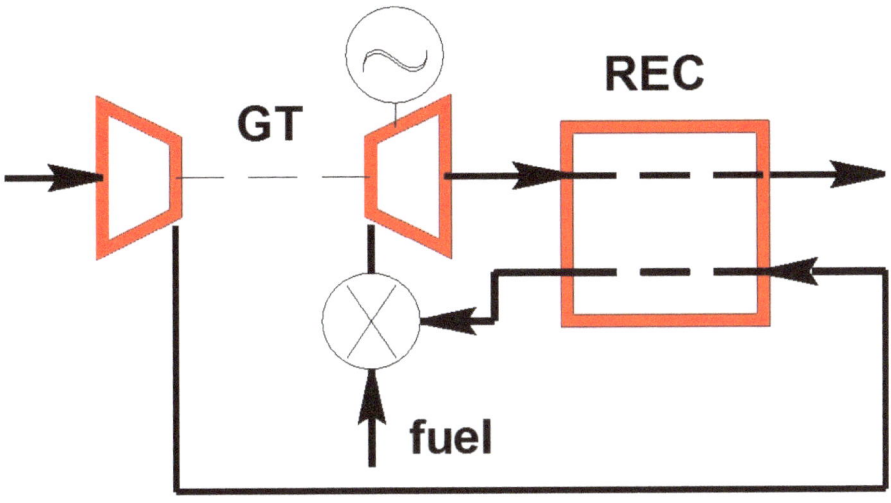

Fig. 9.6

Fig. 9.6 shows a GT cycle with a recuperator heat exchanger (REC) directly heating compressed air entering the combustion chamber. This is another way to improve the cycle's efficiency. This time the hot exhaust services the turbine itself.

Fig. 9.7 shows a liquid natural gas (LNG) production plant directly driven by a gas turbine (GT). Liquefied natural gas is throttled in a valve and fed to the product separator (PS). The LNG product is recovered from the bottom of the separator. Flash vapor from the separator is warmed up in HX, compressed in fuel compressor (FC), and sent to the combustion chamber of the gas turbine (GT). The gas turbine directly drives the refrigerant compressor(s) CMP.

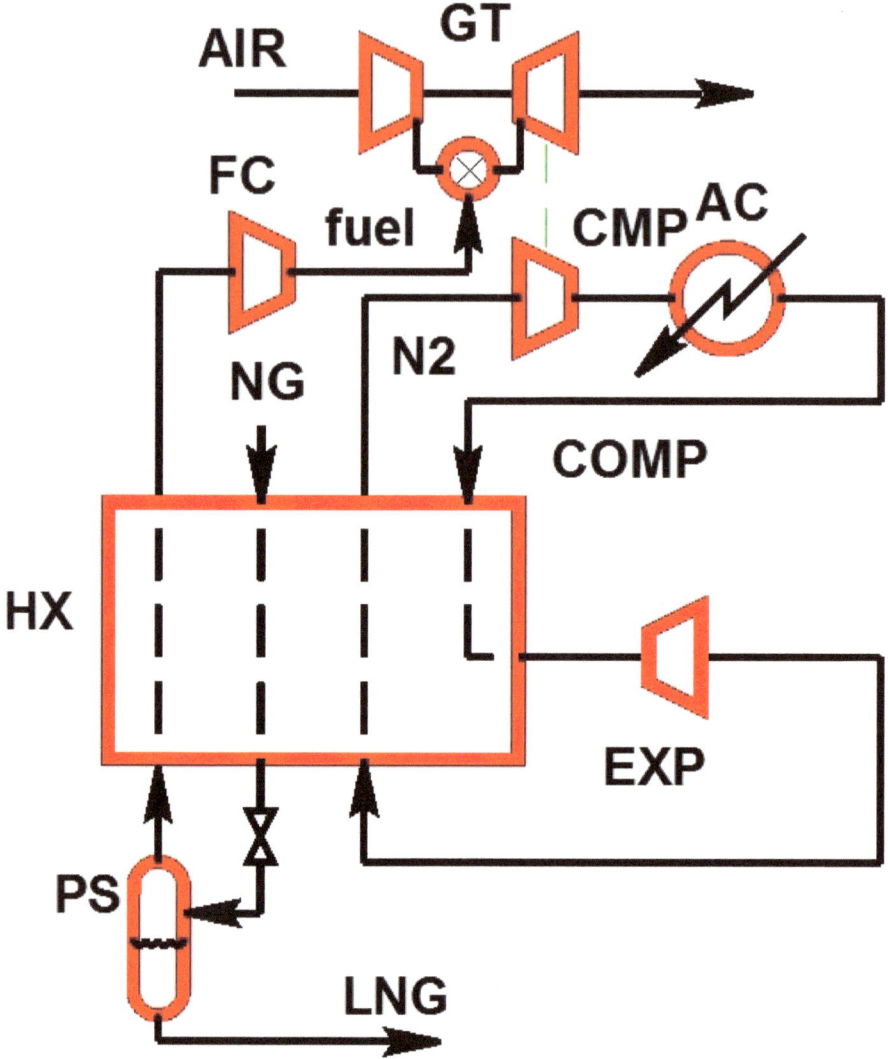

Fig. 9.7

Long before I learned about TRIZ, my colleagues and I were using the Self-Service principle in our engineering practice. We called it colloquially "piggybacking."

This is a good place to explain the patenting process. In order to patent an invention the idea has to be (a) novel; (b) nonobvious (a so-called inventive step once called a flash of genius). For example, a titanium fork may be novel but, since titanium is a metal, and metal forks are known, the question is whether it is obvious to use yet another metal for the same purpose. A feature that sometimes strengthens a patent application is claiming an unexpected benefit.

Before patenting anything, one performs a prior art search. There is usually some prior art concerning most inventions. What it often boils down to is: a combination of feature (a) and (b) is nonobvious in view of (a) and (b) being known as separate entities.

For example, a parabolic solar concentrator is known via the solar power industry. Solar water disinfection (SODIS) is known. Is it nonobvious to combine the two into one invention? What if one claims specific geometry? What if this geometry includes supports that serve multiple functions? Subsequent patent claims become more specific. The first claim may be rejected by the patent examiner, but the subsequent claims may stick. As mentioned before, claiming an auxiliary function strengthens the claim.

UNIVERSALITY AND LATERAL THINKING

This is TRIZ's Principle 6. An object or its parts perform multiple functions; other parts can be eliminated. Like Unification and Substitution, it may follow Subtraction. Unlike Unification, performed functions are interchangeable not simultaneous.

An example of Universality in the natural world is the snowshoe hare. His coat changes from dark to white with changing seasons to avoid predators.

The principle may be related to Reverse Action, but again with functions interchangeable. As mentioned before, an electric motor may become a generator; a pump may become a turbine.

This TRIZ principle is somewhat equivalent to SCAMPER's *P*: Put to Different Use.

I use the principle in at least three ways: (a) to make an object perform different unrelated functions without a modification; (b) to make an object perform similar functions with a modification; (c) to make an object perform different unrelated functions with a modification.

In Jack London's book *The Road*, the author talks about his prison experience. When he was put in jail, he received a loaf of bread. Instead of eating it, he used the wetted inside of the loaf to make a kind of clay to plug the holes wherefrom the bedbugs were coming.

At a museum in Warsaw, Poland, a former Russian prison, I saw a chess set made from bread.

Also I read about somebody escaping from prison by shaping bread into a gun lookalike and painting it black with soot.

My grandmother told me a story about a certain Polish duke who invited to dinner a nobleman known for his wit. Soup was served. All the guests got a spoon, except this one nobleman. "You're a fool if you don't eat the soup," said the duke to everyone. The nobleman cut a heel of bread, removed the soft part, and proceeded to eat his soup with the heel. When he was done, he said, "You're a fool if you don't eat your spoon!"

Those examples are universal uses of bread.

Fig. 10.1 shows an airplane with variable-sweep wing geometry. This produces more lift during takeoff and less resistance at supersonic speed. Similar functions (flight) are performed with a modification to take into account different conditions. As always with Universality, the two functions are not simultaneous.

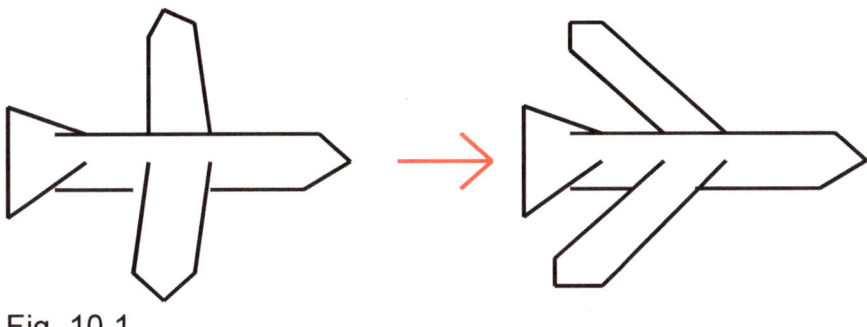

Fig. 10.1

I have my own ruler that also acts as a backscratcher. The completely different functions are performed without a modification.

Figure 10.2 shows a shoe where the high heel can be removed, for example, after a party to walk on the street (US Patent 6,711,835). Similar functions are performed with a modification.

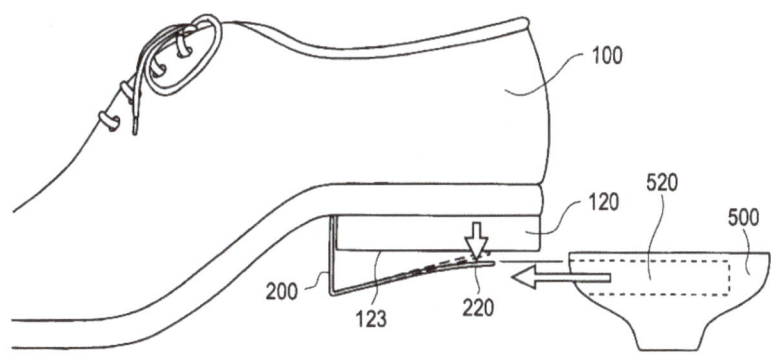

Fig. 10.2

The Universality principle can be applied to classic divergent and lateral-thinking tests. How many uses can one find for a brick? Here are some of the answers I found on the Internet:

    1) To mark the spot where the treasure is buried.
    2) As a unit of measure—could be mass, volume, length.
    3) A stepping stone so one's feet don't get muddy in the yard.
    4) Put it in a toilet cistern to save water (I prefer a plastic bottle).
    5) To draw right angles.

Here are some of the uses I came up with (outside its nominal use to build something):

    1) Zen garden with bricks instead of stones.
    2) Exchange it for money (thugs in Poland would "sell" a brick to a person in exchange for sparing that person's head from being smashed with said brick).
    3) Draw with instead of red paint/crayon (I did it as a child).
    4) To boil water/cook a meal. Brick is, by nature, fireproof. It can be warmed in a fire and used as a heat source.
    5) Paint it to look like a book and put it on a bookshelf.
    6) Support a car on bricks while stealing tires (I have seen it done).

Those various interchangeable uses show a few alternate uses. It is good to have a diverse group of people brainstorm ideas. One should analyze and enumerate the properties of the object (brick: heavy, fireproof, often red, of known dimensions as to weight and volume, edges at right angles, etc.).

I was watching the TV show *Naked and Afraid*. The participants, a man and a woman, are to survive twenty-one days in the wilderness. They are only allowed to take one item each. Many chose a cooking pan or a fire starter. I have a parabolic mirrored Solar Spark Lighter. I also have various camping pots and pans. What if we combine a pan with a fire starter?

Fig. 10.3 shows the device I envision. The inside of the pan is a polished parabolic surface. It can be used to concentrate sunlight to light tinder. The resulting fire can be used to boil drinking water or cook food.

Fig. 10.3

PRIOR ACTION

As I mentioned before, TRIZ principles of Prior Counteraction, Prior Action, and Cushion in Advance, principles 9, 10, and 11, are similar and related to SIT's Division. One can rearrange a process comprising steps A-B-C into, say, B-A-C, to achieve a benefit. The difference is that, with Prior Action, the first step may be an additional step.

One can buy a French liqueur with a pear inside the bottle (*poire prisonnière*). How did the large pear get into the bottle? Driving through the French countryside, one can see pears "growing" on trees. The apparent albeit impossible steps are: (A) grow a pear; (B) put in inside a bottle; (C) add liqueur. The real steps are: (B) procure a bottle; (A) grow a pear inside; (C) add liqueur.

Driving along Trans-Canada Highway, one can see cannon posts. The cannons are used to trigger avalanches before they become dangerous. The natural steps are: (A) snow piles up; (B) more snow piles up; (C) huge avalanche happens. The modified sequence is: (A) snows piles up; (C) small avalanche is triggered; (B) more snow falls on cleared slope.

Here are examples of Prior Action involving an additional step: pre-cut plastic packages of individual servings of ketchup which can be easily opened.

Common application of Prior Action in process engineering is identifying tie-in point for future plant expansions while constructing the original plant. Instead on drilling or cutting the pipes or vessels to connect additional equipment to in the future, one can add taps and flanges as part of the original design.

TRANSFORMATION OF PROPERTIES

TRIZ talks about parameter transitions—change of temperature and/or concentration, or degree of flexibility—called Transformation of Properties (Principle 35) and Phase Transition (Principle 36).

As a chemical engineer, I would merge those principles and both extend them and narrow them. Transformation of Properties could be phase change/transition (solid, liquid, vapor/gas, plasma), change of temperature, pressure, volume, and concentration.

For example, to effectively remove carbon dioxide from a combustion product, one can use pure oxygen or oxygen-enriched air in a combustion device. This involves changing the concentration of oxygen (from 21 % up to 100 %), while reducing the volumetric flow of combustion products, and, therefore, the size of carbon dioxide sequestration equipment. The technique is sometimes called oxy-fuel.

Natural gas is liquefied and transported across the ocean as LNG (liquid natural gas), dramatically reducing the storage space. Phase change results in reduction of volume.

Other examples include a pressure cooker to shorten cooking time at increased pressure and, therefore, temperature; steam cooking of vegetables to preserve vitamins and taste (steam instead of immersion in boiling water); liquid soap from a dispenser instead of bar soap for a more sanitary use at a public bathroom; superconductivity at low temperature; and supercritical CO_2 for extraction and for silicon wafer cleaning.

Let us look again at the fish pond example. Oxygen concentration in a fish pond has to be increased from one thousand ppm to two thousand ppm for the optimal conditions in which to breed the fish. The common solution is to compress air and inject it into the pond through a perforated pipe. Here are other solutions that use the Transformation of Properties principle.

Fig. 11.1 shows the solution suggested in *40 Principles: TRIZ Keys to Technical Innovation* by Genrich Altshuller. Gases are more soluble at higher pressure. Water can be saturated with air at increased pressure and then pumped into the pond.

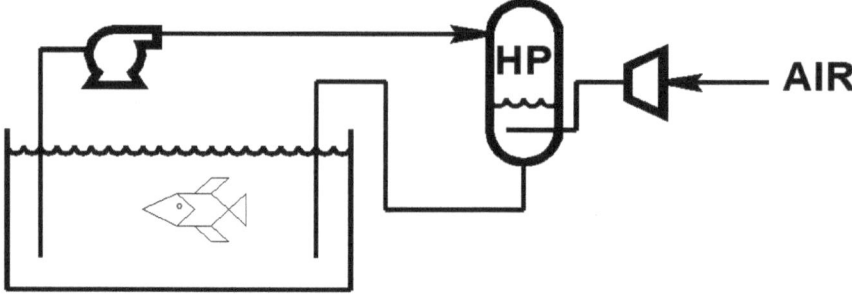

Fig. 11.1

Fig. 11.2 involves a temperature change. Gases are more soluble at lower temperatures. Water can be chilled, saturated with air, and then pumped into the pond. This solution also goes back to Prior Action described in a previous section.

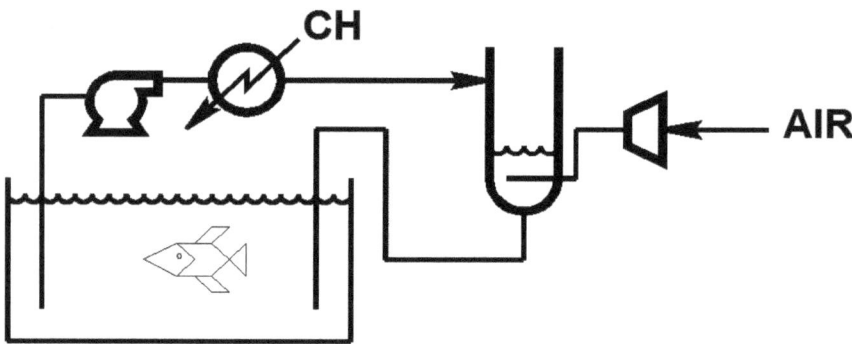

Fig. 11.2

Fig. 11.3 shows a different solution. One can change the concentration. Pure gaseous oxygen (GOX) or oxygen-enriched air can be used instead of plain air.

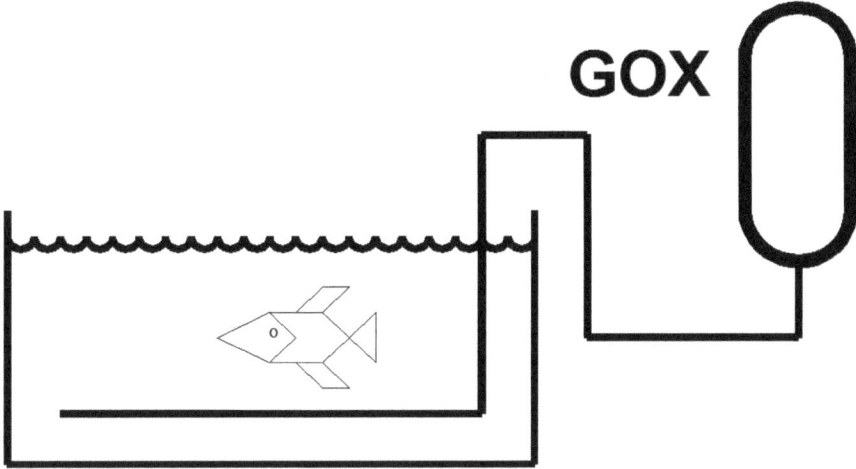

Fig. 11.3

One can also use liquid oxygen (LOX, Fig. 11.4). It will simultaneously decrease temperature and supply pure oxygen. LOX should evaporate quickly against ambient heat sink.

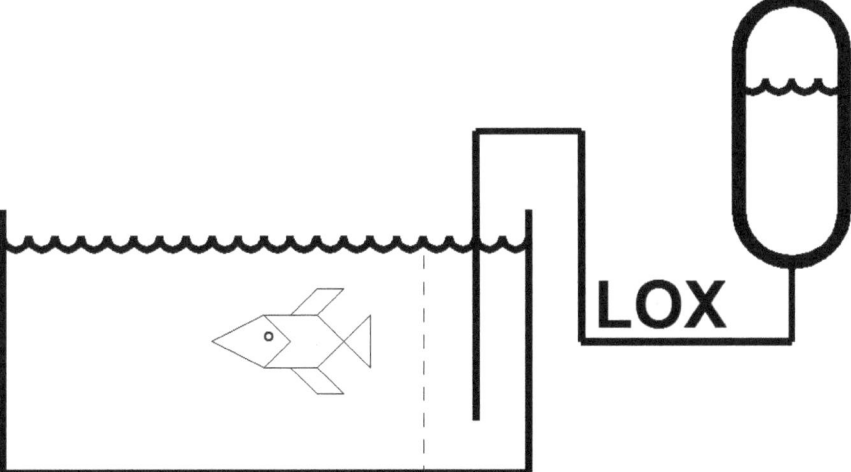

Fig. 11.4

Fig. 11.5 shows a more complex example of phase change, a cryogenic ASU (air separation unit). Gaseous oxygen is withdrawn from the bottom of the low-pressure column (LPC), warmed up in the heat exchanger (HX), and compressed to the desired pressure in the compressor (CMP).

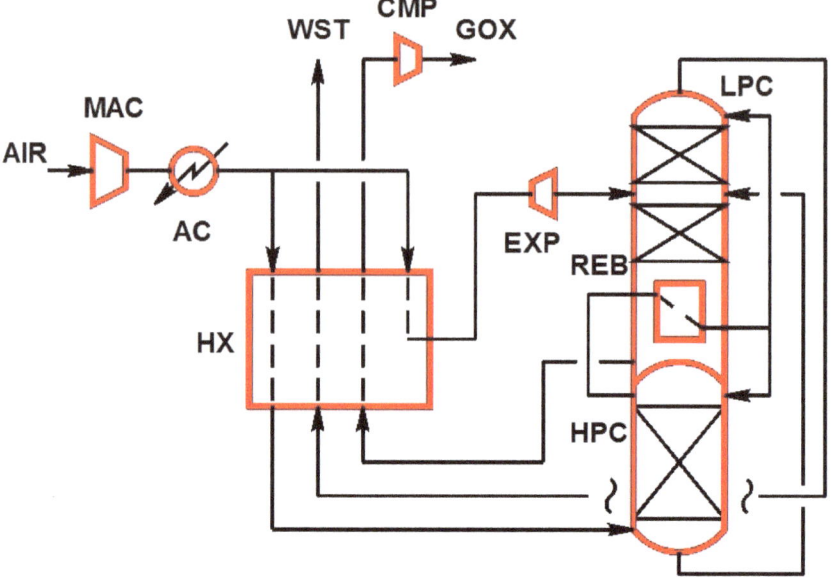

Fig. 11.5

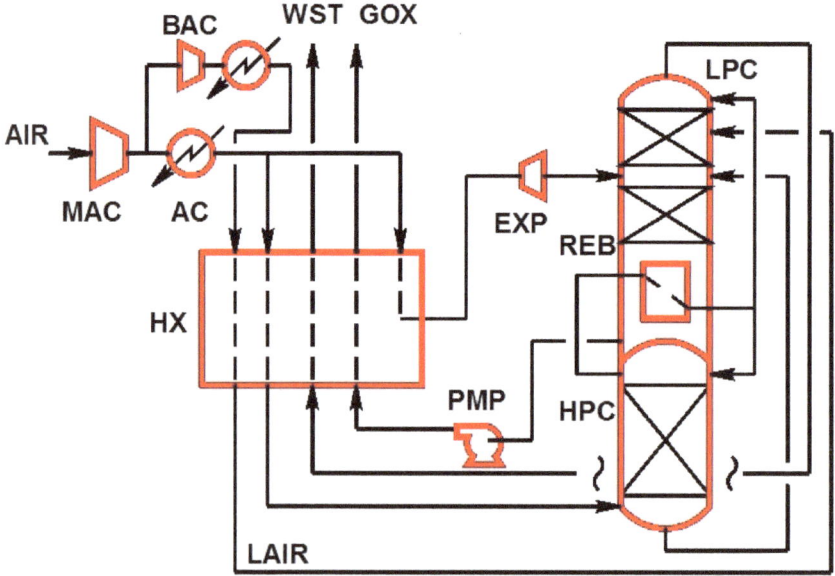

Fig. 11.6

Fig. 11.6 shows a different solution, a so-called pumped LOX (liquid oxygen) cycle. LOX is withdrawn from the bottom of the low-pressure column and pumped to the desired pressure in the pump (PMP). Pressurized LOX is then vaporized in the heat exchange (HX). Pumping LOX versus compressing it is a trade-off between the pump cost versus the compressor cost.

In order to accommodate this additional latent heat duty, a portion of air is compressed to a higher pressure in the booster air compressor (BAC) and condensed in the HX into liquid air. So, effectively, a potentially more expensive oxygen compressor (CMP) may be replaced with a conventional air compressor (BAC).

Here is a field example from Engineers Without Borders in El Salvador. We could not buy PVC elbows for water piping other than at ninety-degree or forty-five-degree angles. Those were not adequate for connecting the water tank with the valve box. So we heated a 4" PVC pipe over an open fire normally used for cooking, bent it, and cooled it back down with water.

This technique involved changing the degree of flexibility by changing temperature. We achieved a snug fit.

I recently faced the following dilemma. The NSA regulations do not allow large quantities of liquids to be taken on board commercial airplanes. In addition, one cannot take on board airliners pressurized containers. On top of that, airlines now charge fees for checked-in luggage. I was going to a tropical country for two weeks with only a carry-on. I needed enough mosquito spray for fourteen days.

I bought a small bottle containing 98 % DEET in a size allowed by the NSA. I also bought a larger empty bottle with a spray pump (not a pressurized container). Upon arrival at my destination, I diluted the DEET with alcohol (DEET is insoluble in water) to get three times as much spray with about 30 % concentration, perfect for mosquito protection and safe for the skin.

TRANSITION TO A NEW DIMENSION

This is TRIZ Principle 17. We can learn about it from nature. For example frogs' and hippos' eyes have migrated to the top of their heads to allow for improved vision while having most of their bodies submerged in water.

Let is consider the following puzzle. We have six matchsticks. The task is to build four equilateral triangles, no more, no less.

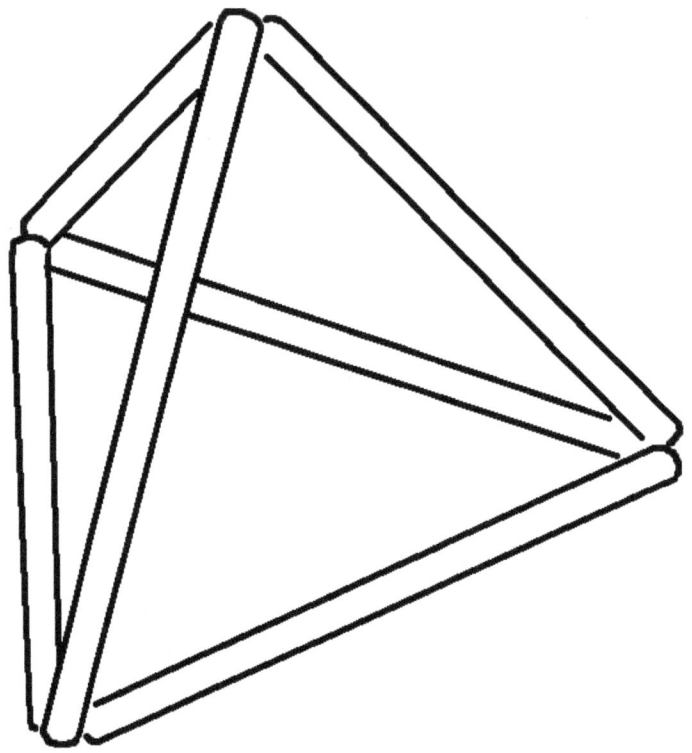

Fig. 12.1

Fig. 12.1 shows the solution. Most people first seek a solution in two dimensions. Only some realize that, by introducing the third dimension, the solution presents itself.

Fig. 12.2 shows another classic puzzle: how many triangles do you see?

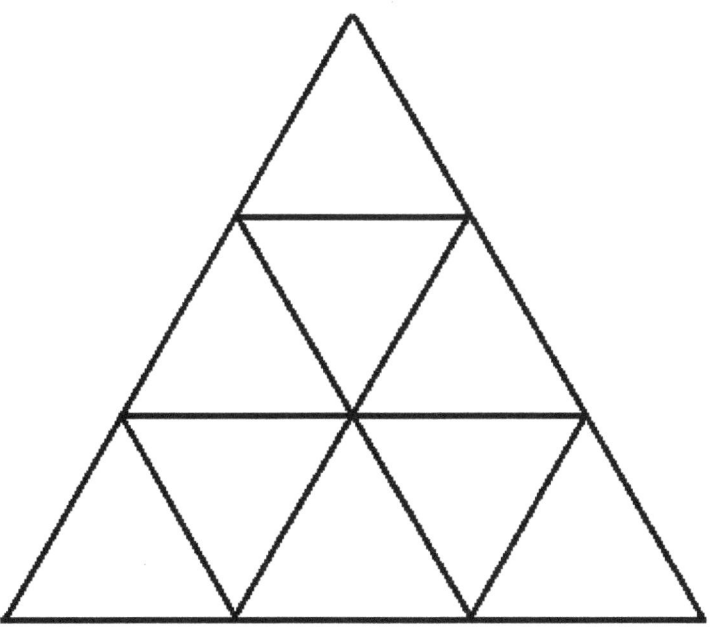

Fig. 12.2

The conventional answer is thirteen. What if we bring the solution to a new dimension, this time by assuming that we are looking at the object from above (Fig. 12.3)? Now the answer becomes nine that one can see from above, and at least fifteen if one counts the vertical surfaces twice, on the outside and on the inside.

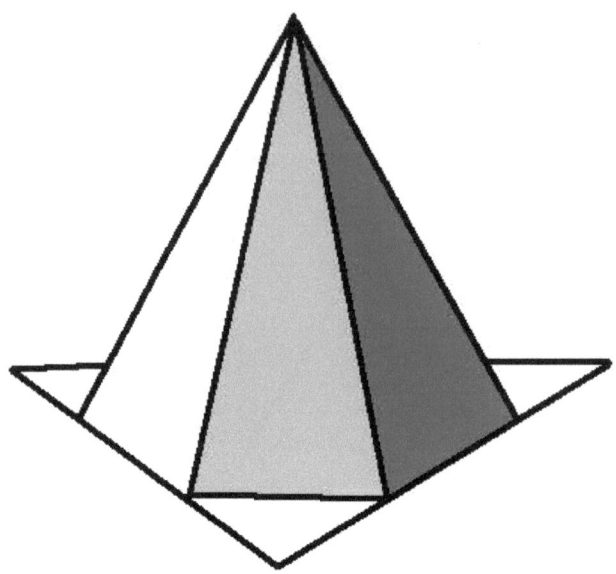

Fig. 12.3

The Transition into a New Dimension principle can be used (a) to improve an existing system/process/solution; (b) to come up with a new application or invention.

Fig. 4.1 shows a parachute used to save a falling plane, a nominal use of a parachute with a person replaced by a plane (Substitution). What if one makes a parachute horizontal instead of vertical (Transition into a New Dimension)? Fig. 12.4 shows a parachute used to slow down a landing plane.

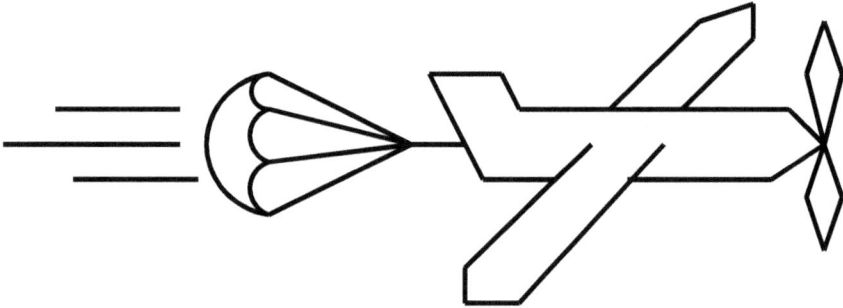

Fig. 12.4

What it we move a propeller from the top of the helicopter to the bottom? Sounds like a crazy idea?

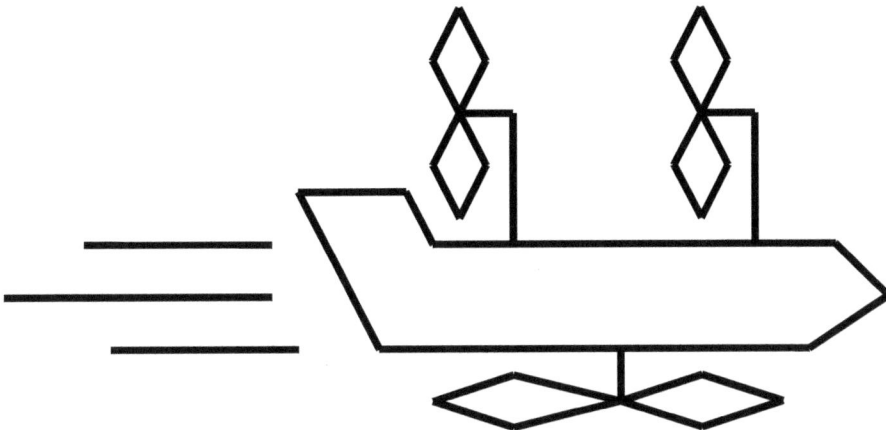

Fig. 12.5

We have a new invention, a hovercraft (Fig. 12.5)!

A classic example of the use of the Transition into a New Dimension principle in engineering is taking advantage of so-called critical elevations. In the section on Subtraction, I mentioned a car with the fuel tank located above the engine, no fuel pump. In the section on Multiplication, I mentioned a water tank at an intermediate elevation to save power.

Transition to a new dimension may mean reducing the number of dimensions. I have a SwissCard, a flat, essentially two-dimensional version of the Swiss Army Knife the size of a credit card. I also have a flat magnifying glass using the Fresnel lens. Lightweight Fresnel lenses were once a breakthrough in lighthouse construction.

REPLACEMENT OF A MECHANICAL SYSTEM

This TRIZ Principle 28 is as important today as it was when Altshuller first started working on his system. In process engineering, we are often trying to make the process as efficient as possible. Sometimes it affects reliability. The more moving parts, the more likely the system or the device will fail. For example, the vortex tube shown in Fig. 7.4 generates refrigeration without any moving parts (at the cost of efficiency).

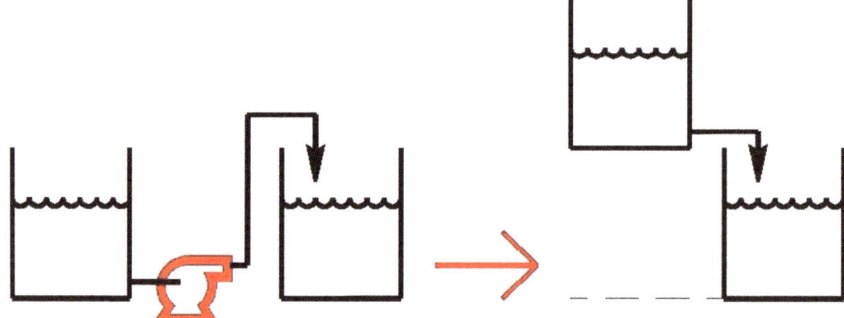

Fig. 13.1

Fig. 13.1 shows a solution that illustrates the use of critical elevations mentioned in a previous section. The mechanical system (pump) is eliminated by elevating one of the tanks and using gravity to fill the other.

PARTIAL BENEFIT AND TAKING TO EXTREME

TRIZ Principle 16, Partial or Excessive Action, says that, if it is difficult to achieve a certain outcome, one should try to achieve more or less.

This principle was present in nature long before TRIZ. Butterflies are trying to avoid predators. Some evolved to have drab colors to merge with the background. Others—for example, the monarchs—developed very bright colors to warn predators about their toxicity or foul taste. Those two approaches are the opposite but they both work to the same end.

The Pareto principle, sometimes called the 80-20 Rule, implies, among other things, that one can achieve 80 % of the benefit from 20 % of the effort.

What I often see in my engineering practice is that a problem may not justify the additional effort or expense. In the chemical industry we are often asked for the four Ps: power (consumed or generated), production, purity, and, in the case of industrial gases, pressure. Each time one should ask the question whether those numbers make sense for a particular application. Do we need the oxygen 99.9 % pure for a fish pond? Do we need it at high pressure for a sanitation project?

Some examples of Partial Benefit were shown in the prior sections on Extraction and Unification. For example, the distillation column in Fig. 2.6 may be less efficient than the one in Fig. 2.11 but the additional efficiency may not be justified by the product purity requirements. The ASU in Fig. 2.15 makes only nitrogen; the one in Fig. 2.11 makes both oxygen and nitrogen. The client may not need oxygen production.

Conversely some systems and processes can easily be modified for an additional benefit.

Classic TRIZ postulates that one should not accept a compromise, a partial solution. I always look at a possibility of obtaining Partial Benefit. It is a very important concept and naturally follows Subtraction and Unification. One can apply Reverse Action and/or Convert Harm to Benefit principles to find an application that fully takes advantage of the process or system that offers Partial Benefit.

Sometimes it is possible to get 100 % benefit for 50 % or less of the effort.

$$z_{i+1} = z_i{}^4 + c$$

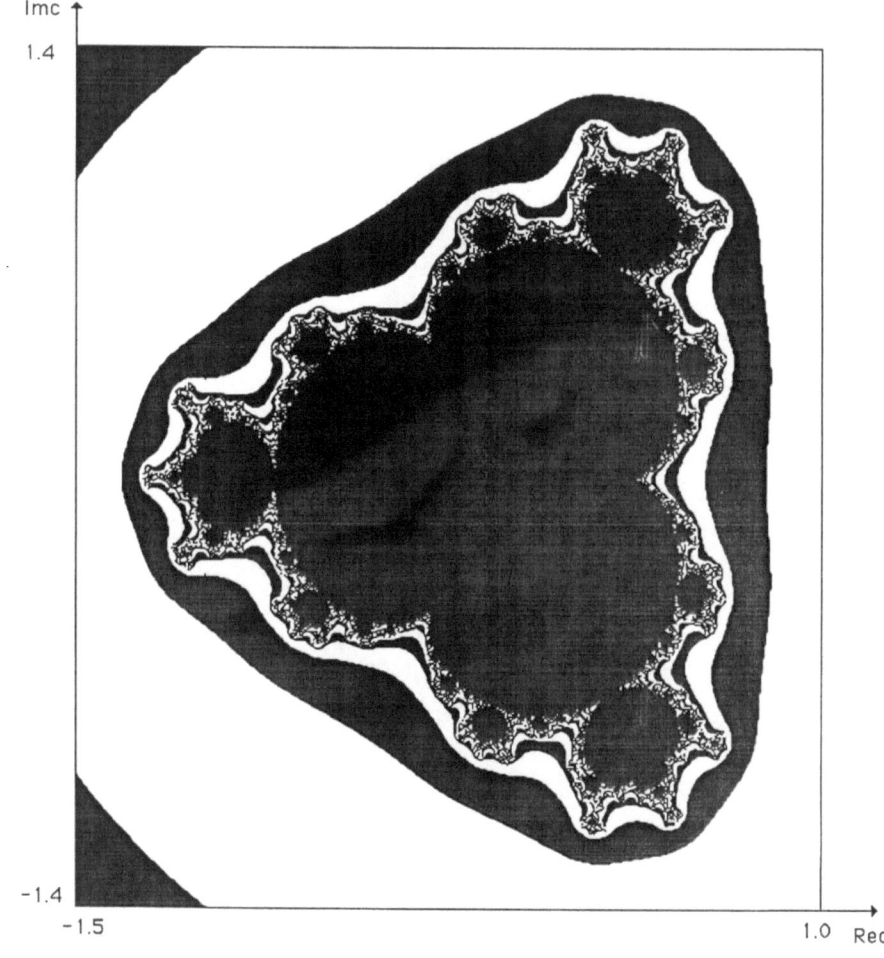

Fig. 14.1

Some twenty-five years ago a fellow student with whom I had an internship wrote a program that produced a Julia set and a Mandelbrot set. The equation for the Mandelbrot set is:

$$Z_i + 1 = Z_i \char`\^ 2 + C$$

Z and C are complex numbers. I modified his program to produce something else. First I tried raising the complex variable Z to the third power. The program ran all night on the old Macintosh PC, and the result did not look good. Then I tried the fourth power. It looked beautiful! The program terminated early, so the image was not complete, but I knew it had to be symmetrical. I used a graphics program to generate a mirror image of the completed part of the fractal and produced the image shown in Fig. 14.1.

Therefore, I produced "my own" fractal for 50 % of the computational effort, taking advantage of a certain property. In retrospect, the idea was fairly obvious. I later found "my" fractal as an option on the commercial fractal program.

To me, TRIZ Principle 16 (Partial or Excessive Action) suggests a general problem-solving technique that involves Taking It (the problem) to the Extreme, sometimes two opposite extremes, to bracket the answer and to find a solution or a trend.

When I was in high school, my math teacher explained to us how to divide an angle into two equal angles using a straightedge and a compass. He also explained how to divide a line segment into any number of equal segments. Then he asked us to divide an angle into three equal parts (an angle trisection problem). I thought I had a solution. I would connect the relatively sharp (acute) angle he drew with a line segment (a chord), divide the segment into three parts, and that way divide the angle. I did not give it too much thought. The teacher told us triumphantly that it was not possible to divide an angle into three equal parts by only using a straightedge and a compass. He left it at that.

Let us Take It to the Extreme (Fig. 14.2). As the angle approaches 180 degrees (an obtuse angle), one can see that, out of the three angles created by equally dividing a line segment, the two outer angles are equal, and the middle angle is bigger. When the angle becomes 180 degrees, the two equal angles become zero and the middle one becomes 180 degrees as well. One can see clearly that the proposed method does not work.

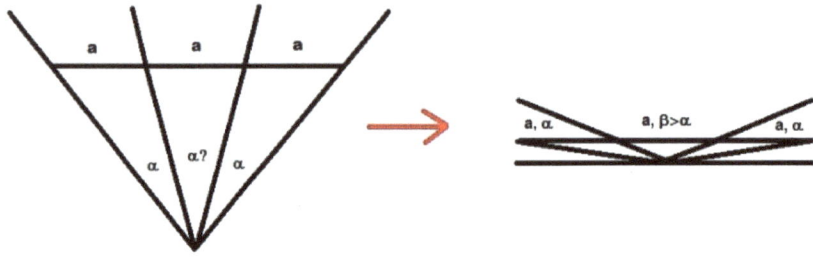

Fig. 14.2

Let us consider the following problem. We have two identical drinking glasses, the first full of alcohol and the other one full of water. We take a spoonful of alcohol from the first glass, put it in the second glass, and carefully stir it. Then we take a spoonful of the resulting mixture and put it in the first glass. Do we have more water in the first glass or more alcohol in the second glass?

There is more than one way to solve this puzzle. Let us bracket the answer. Let us assume that the spoon is infinitesimally small. If we use such a spoon, nothing happens. We have the same amount of water in the first glass and alcohol in the second glass, namely zero change. Let us now assume the spoon is huge and will contain all the liquid from each glass. Then obviously we have equal amounts of water and alcohol in each glass. Through interpolation, the correct answer is the amount of water in the first glass is equal to the amount of alcohol in the second glass.

Here is another puzzle, a question supposedly asked at job interviews. If one does not have an umbrella, is it better to run in the rain to reach the car in the parking lot or is it better to walk slowly? Let us assume we walk at near-zero speed.

We will obviously get soaked. What if we instantaneously teleport? We will remain bone-dry. So here is the answer.

An airplane flies from point A to point B at 600 mph. It has a 50 mph tailwind. When it flies back, it is flying against the 50 mph wind. Will the tailwind affect the round-trip flight time? Any tailwind will accelerate the flight. What if the wind speed is 600 mph, unrealistic but useful as a thought experiment? Then the airplane will not move at all on the way back! Any wind will lengthen the round-trip flying time.

Fig 14.3

The famous Monty Hall problem requires the participant to choose one of the three doors (Fig. 14.3). The participant knows a car is behind one of the doors and a goat is behind each of the two remaining doors. The host, who knows what is behind each door, opens one of the remaining doors to reveal a goat. Is it to the participant's advantage (assuming they want a car not a goat), to switch to the remaining door? Marilyn Vos Savant implicitly used Taking It to the Extreme technique to reformulate the problem. Supposing there are one hundred doors. The participant chooses one. The host, who knows what is behind each door, opens 98 doors to reveal 98 goats. By now the answer should be obvious.

I successfully use the Taking It to the Extreme technique to solve scientific and engineering problems, and to determine trends in various "what if?" scenarios.

IDEAL FINAL RESULT

If you want to build a ship, don't drum up the men to gather wood, divide the work and give orders. Instead, teach them to yearn for the vast and endless sea.
—Antoine de Saint-Exupery, writer, pilot, pacifist

If you have built castles in the air, your work need not be lost; that is where they should be. Now put foundations under them.
—Henry David Thoreau, writer, philosopher

Measure strength for your goals, not (choose) your goal according to your strength.
—Adam Mickiewicz, nineteenth century Polish romantic poet

A problem well-stated is a problem half-solved.
—Albert Einstein.

Ideal Final Result (IFR) is a powerful foundational TRIZ concept that describes the optimal solution to a problem, without jargon (jargon tends to limit the solutions to a certain domain) and independent of the original constraints. The TRIZ definition of the ideal system is that it occupies no space, has no weight, requires no labor, uses no energy, requires no maintenance, and/or costs nothing. My definition is simply the best theoretical solution.

IFR has the following four characteristics:
1) It eliminates the deficiencies of the original system.
2) It preserves the advantages of the original system.
3) It does not make the system more complicated (uses free or available resources).
4) It does not include new disadvantages.

IFR may be hard or even impossible to achieve but it (a) encourages breakthrough thinking and overcomes psychological inertia; and (b) establishes the target, the idealized solution to aim for, or a benchmark to compare against.

Defining IFR may include estimating the maximum possible benefit that can be derived from an unconstrained solution to the problem. Constraints may be imposed later.

When Heracles (Hercules) took up the task of cleaning the Augean stables in a single day, he did not pick up a shovel or a bucket. He first looked for available resources. The available resources were nearby rivers. He made two openings in opposing stable walls and diverted the rivers to flow through the stables (Fig. 15.1). This took care of the problem. His IFR was clean stables, not the best or the quickest way to shovel the dung or to carry it on one's back. The time constraint actually helped because, given more time, he could have come up with a more conventional solution. I will talk about imposing constraints later in the book.

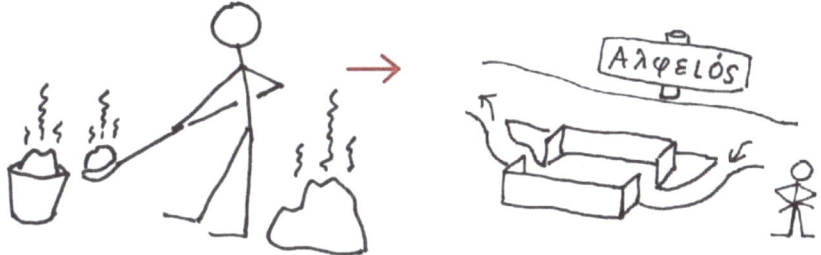

Fig. 15.1

At some point I read about an efficient stove that uses less wood to cook food in developing countries. It was a quite fancy thermal design. It was supposed to be deployed to a hot desert location. I pointed out to the designer that one could use the solar cooker instead (Fig. 15.2). It uses zero fuel. It takes advantage of the abundant resource and, for all practical purposes, an infinite source of energy. Of course, the efficient stove has its applications, for example, at night or during the rainy season. The IFR here was not an efficient stove or less fuel consumption but getting food cooked, ideally using no fuel.

Fig. 15.2

I often use the concept of IFR in my process engineering practice, especially in thermodynamics.

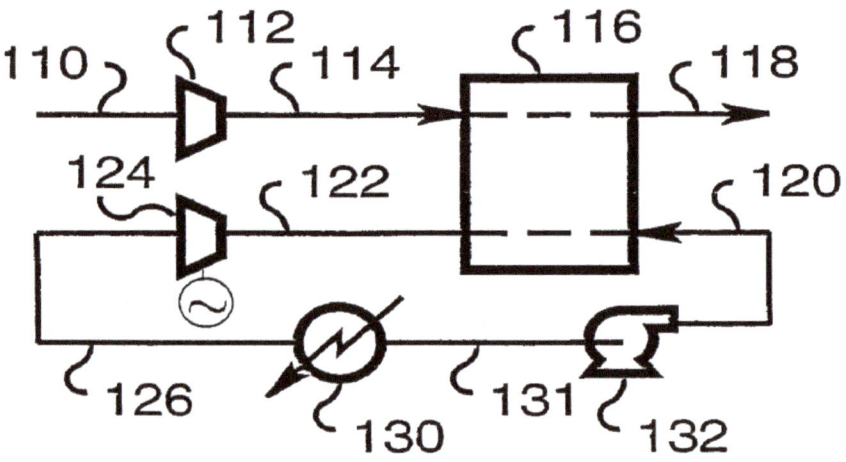

Fig. 15.3

When gas is compressed, part of compressor's work shows up as increased pressure of the discharge stream. Another part is wasted as heat. An IFR would be for all compressor work to increase the pressure, with none wasted as heat. So how much work is wasted as heat? Exergy analysis is done on an air compressor with the typical compression ratio of 1.8.

$$\text{Exergy} = \Delta (H - To\ S)$$

where H is the enthalpy, S is the entropy, and To is the reference (standard) temperature. The calculation indicates that, at certain conditions, about 10 % of the compressor's work is wasted as heat. Therefore, up to 10 % of the compressor's work can be recovered. Now that we know our target, we can (a) decide if we really want to recover this energy; (b) look at existing solutions to similar problems; (c) devise a scheme; (d) evaluate its efficiency.

Fig. 15.4 is from US Patent 7,278,264 (I am the inventor) and shows one solution to the problem. Dense fluid is heated against the hot compressor discharge in a heat recovery exchanger and expanded in a two-phase dense fluid expander (hydraulic turbine) to generate power. Six percent of compressor's work is recovered as power generated by the turbine minus the pump's work (60 % of the IFR).

In engineering practice we often want to establish an ideal benchmark to compare against. We have to make sure this is the absolute best solution. We often see unintentional, and sometimes intentional, comparisons against a straw man device or process.

In thermodynamics and process engineering, we often want to establish the absolute minimum energy consumption of a process that consumes energy or the maximum energy that can be produced or exported. A lot of emphasis nowadays is on environmental aspects. One has to bear in mind that the cleanest energy is energy not used.

OLD SOLUTIONS AND SOLUTIONS FROM A DIFFERENT DOMAIN

> "This is nothing, this is nothing," he was saying touching the wound with his fingers. "He is going to be fine tomorrow. I will take care of it. Mix the spiderweb with bread."
> —Henryk Sienkiewicz, *The Deluge*, Volume II, 1886.

In no less than three places in his trilogy, written in the late nineteenth century, the Nobel Prize–winning Polish author Henryk Sienkiewicz mentions using a mixture of bread and spiderwebs to dress wounds. He talks about events taking place in Poland in the seventeenth century, based on old writings from that era. Why would anyone use a mixture of bread and spiderwebs?

Penicillin was serendipitously discovered by another Nobel laureate, Alexander Fleming, in 1928. It could have been discovered much earlier if somebody had asked the question, why does a mixture of bread and spiderweb work? It works because the web contains sticky silk that captures the *Penicillium* spores. The bread provides nutrition for the fungus to grow.

While visiting the Jha-Karpo (White Cliff) monastery near Paro, Bhutan, I saw a prayer wheel moved by hot air. I was not allowed to take photos at the monastery so I did a sketch. Fig. 16.1 shows the original sketch from my travel notebook and the subsequent patent drawing from US Patent 6,135,603.

I have a book published in 1948, entitled *2100 Needed Inventions* by Raymond F. Yates. It provides lists of needed inventions in various disciplines. It is interesting to see how things have changed since he wrote this. Many inventions have become a reality. But others have not. The book is a good exercise in idea-generation. For example, General Problem number 558 is a nontipping ink bottle. I thought of a roly-poly doll, such as the Russian *vanka-vstanka*, one that rights itself when pushed over. One could design similar bottles and other liquid vessels with semispherical weighted bottoms. As I expected, somebody already patented the Self-Righting Bottle (US Patent 6,168,034; Fig. 16.2). There was a lot of other prior art.

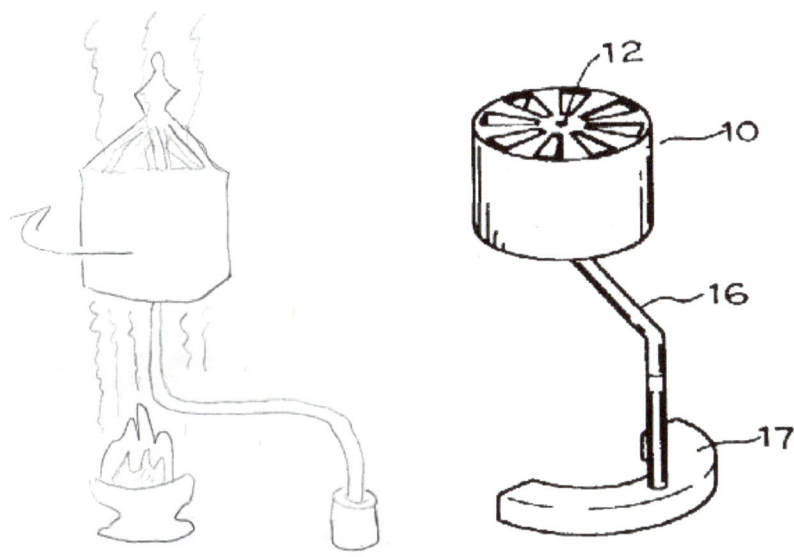

Fig. 16.1

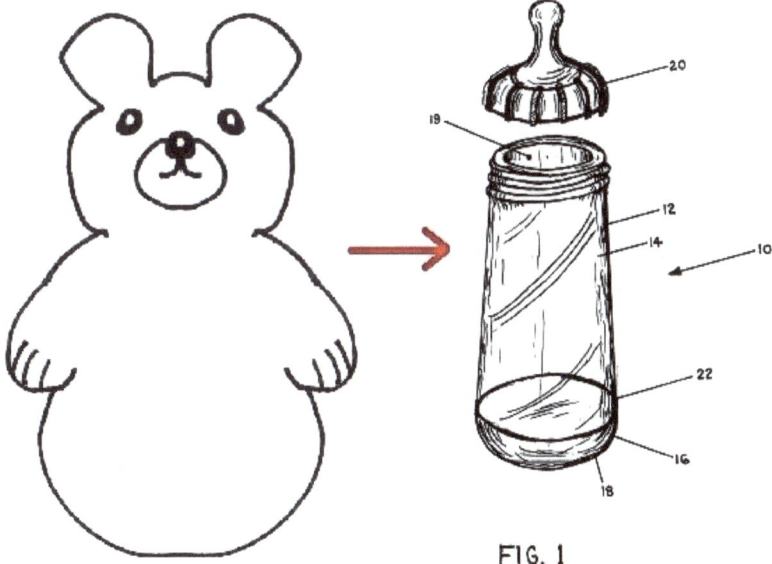

Fig. 16.2

When I visited Venice, I found out the seafaring Venetians built their ships using a method very similar to the modern assembly line as early as the fourteenth century. Yet Henry Ford is credited with development of an assembly line in the early twentieth century. He allegedly got the idea after visiting a meat packaging plant. This is another example of transferring knowledge from one discipline to another. Modern industrialists could have gotten this idea earlier had they studied the Venetian shipbuilding techniques.

Fig. 16.3

TRIZ postulates that somebody has already solved your problem or a similar problem. This usually involves relatively recent developments. But there is so much to rediscover by studying ancient sources and old cultures. In the Middle Ages people thought the Earth was flat, but the Greeks had measured its circumference much earlier.

The prayer wheel patent is an example of an old solution of the same problem. The self-righting bottle patent and the Venetian assembly line are examples of applying old solutions to different disciplines.

This brings us to the concept of The Enabler. Some of the inventions and discoveries were known for a long time. They did not find application at the time they were invented or discovered because there was no application or no technology to make them practical. But they may find an application today. I call the breakthrough that can make them practical

The Enabler. An excellent example is the fuel cell. The first fuel cell was invented in 1838. But it was not until recently that fuel cells found a practical application.

As mentioned before, penicillin was unknowingly used for centuries. It was officially discovered in 1928. But it was not until the 1940s that it was produced on an industrial scale at high-enough quantities to make a difference in fighting diseases.

The 1942 US Patent 2,292,387, entitled Secret Communication System but called the "Frequency Hopper" (Fig. 16.4), coinvented by the famous actress Hedy Lamarr, had not become practical until the 1950s and is now appreciated for its use in modern cell phones.

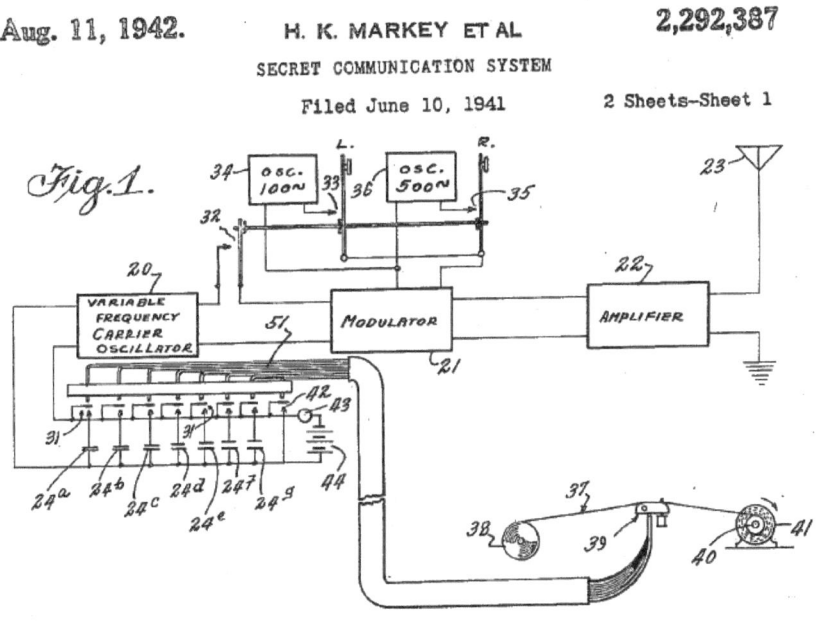

Fig. 16.4

Remember the bread spoon example? It is not that crazy. I have a set of biodegradable utensils made from starch.

I enjoy reading old books on science and technology for inspiration of new inventions to create and for application of some Old World ideas into a new market.

CONSTRAINED CREATIVITY AND THE WORLD CLOSED PRINCIPLE

> The more constraints one imposes, the more one frees one's self. And the arbitrariness of the constraint serves only to obtain precision of execution.
> —Igor Stravinsky

> Use it up, wear it out, make it do, or do without.
> —Amish proverb

Goffin's cockatoos (*Cacatua goffiniana*) are some of the most creative birds, producing makeshift tools to solve problems. For example, they may modify wood to make a tool to retrieve food. One theory explaining their creativity is that they evolved on small islands with limited resources. Nature teaches us that imposing constraints may actually increase creativity. This lesson is much older than any systematic innovation tool.

"Étude Op. 10, No. 5, in G-flat major" was composed by Frédéric Chopin in 1830. The accompaniment is played by the right hand exclusively on the black keys of a piano. Limiting it to black keys makes it more interesting.

A lipogram is a text written by omitting a certain letter. The most common letter in the English alphabet is *E*. *Gadsby* is a 1939 novel by Ernest Vincent Wright that does not include words that contain the letter *E*. This is so-called "constrained writing."

Avoiding a particular building block may spawn many novel applications. Notice I did not use the letter *E* in the last sentence.

Fig. 17.1 shows the classic nine-dot puzzle used by various creativity gurus presenting "thinking outside the box," often as part of employee training. The task is to connect all nine dots with straight lines without lifting the pencil.

I am a big fan of thinking outside the box. For the time being, let us concentrate on what is inside. The Closed World principle (seeking solutions within a certain domain) is one of the basic SIT concepts.

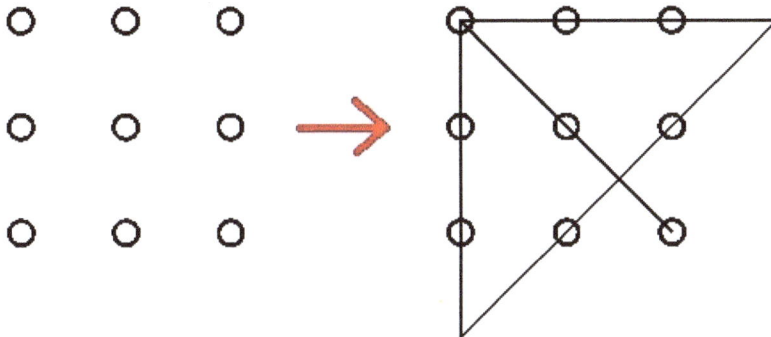

Fig. 17.1

Duncker's candle problem was a cognitive performance test first administered in 1945. Someone is given a tapered candle, a box of matches, and a box of thumbtacks (Fig. 17.2). The task is to attach the lit candle to the wall so that the wax does not drip on the table below. Some people try to attach the candle to the wall with thumbtacks. Others use dripping wax as an adhesive. Those solutions do not work too well.

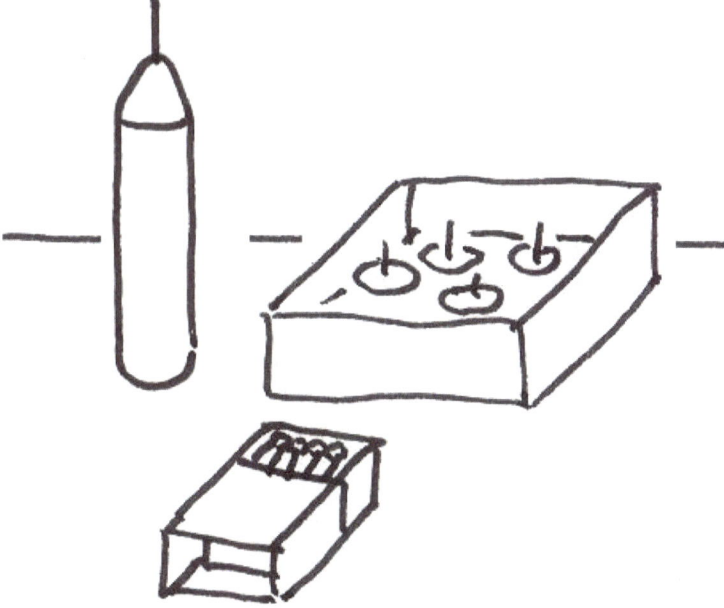

Fig. 17.2

The preferred solution is shown in Fig. 17.3. The box that contained thumbtacks is attached to the wall to support the candle.

Fig. 17.3

Fig. 17.4

It turns out people are more likely to solve the puzzle if the tacks are shown beside the box (Fig. 17.4). This breaks "functional fixedness." We tend to assume that certain objects fulfill certain functions, and we reluctantly use them for other purposes. SCAMPER's *P* principle (Put to Different Use) could also lead to the preferred solution. Are there any other solutions?

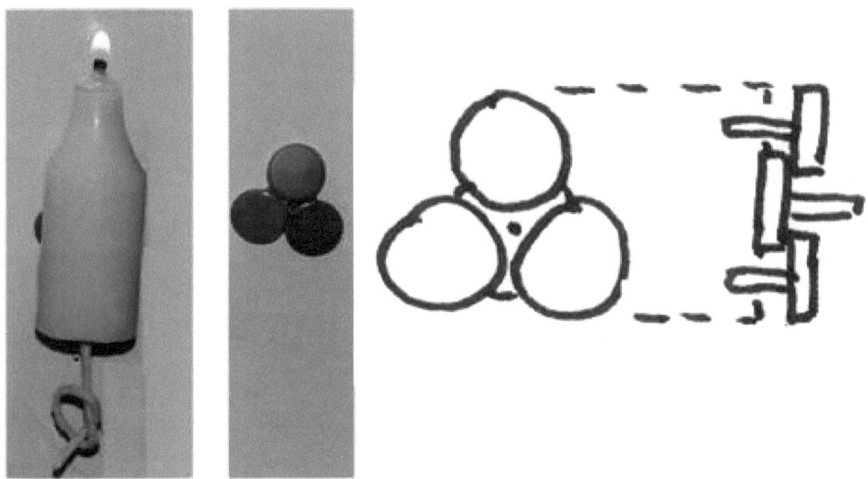

Fig. 17.5

What if the box is not sturdy enough? I had another idea, shown in Fig. 17.5. With three other thumbtacks, I attached one reversed thumbtack to the wall with its pin sticking out. Now I could easily attach the candle to the wall.

I also stripped part of the candle to expose the wick. The wick is a string. This opens another realm of possibilities. Do you have any suggestions?

I used Division to divide the system into components: wax, wick, tacks, box, and matches. Now I can use any subset of the components to solve the problem without introducing additional components.

Another way to look at this problem, and many others, is to identify the domain to play within. The red box bordering the whole of Fig. 17.6 is the original domain. The green box allows solutions involving only thumbtacks and the candle. The pink box only allows solutions involving the candle itself (wax, wick, etc.). The blue box, shown as outside the red box but open on the right-hand side, can extend the domain to other solutions. No matter what we do, we always work within some closed domain!

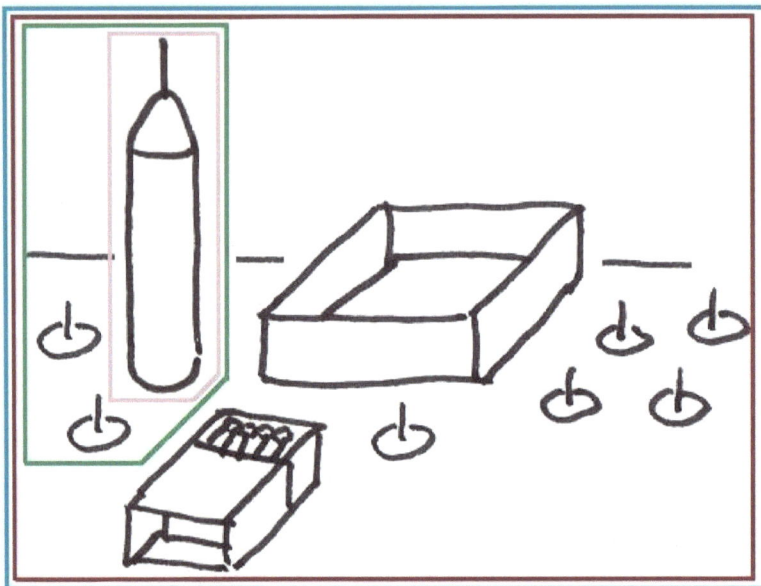

Fig. 17.6

The Closed World principle is applied implicitly in jury-rigging (also called jerry-rigging, improvising solutions). I have this elementary school friend who was a real-life MacGyver before there was the TV series *MacGyver*. I had this pencil case with a zipper. The zipper pull tab broke off. I would break my fingernails operating the zipper. He took a piper clip and inserted it into the slider to produce a new pull tab.

We went on long bike rides across the countryside in our custom-modified bikes. Sometimes the bearings in one of the pedals would break. He would take some butter from a sandwich and put it in the bearing to provide temporary lubrication.

In Jules Verne's *The Mysterious Island* the protagonist makes a lens by using two curving glass surfaces obtained from watches, sealing them together, and filling the space between them with water. He then uses the lens to concentrate sun rays to make fire. I repeated this experiment by removing two lenses from reading glasses, putting them together using putty (one could use resin from a conifer tree), and filling the space in between the lenses with water.

I set an oak leaf on fire (Fig. 17.7). A single reading lens would not have done it. Here I applied SIT's Closed World principle: I used only the articles I had on me during a hike in the wilderness, including the glasses to read the map at dark. I also used TRIZ's Mediator principle. I used water as the mediator with its high refractive index to connect the lenses.

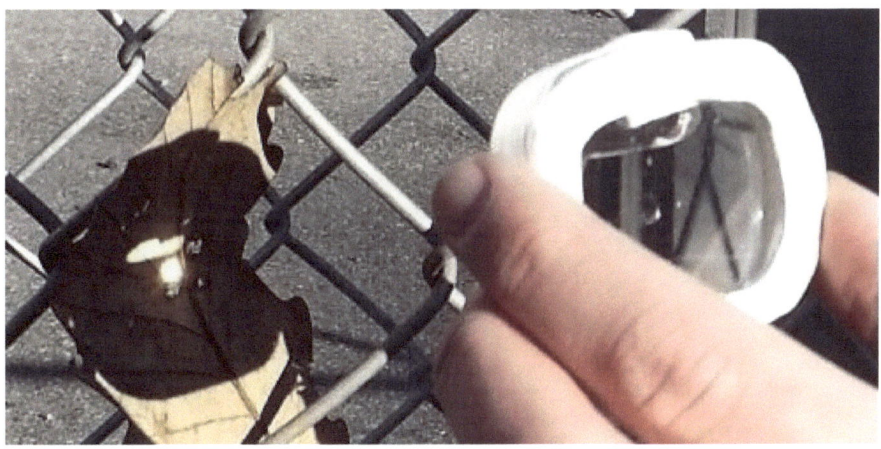

Fig. 17.7

I used further constraints when I drew most illustrations in this book by using a simple CAD–like program that allows drawing only straight lines, circles, and ovals. This actually made me more creative when it came to drawing, for example, an airplane. I mostly used two colors: black and red.

Plus, in writing this book, I constrained myself to about one hundred pages, give or take a few. This helped me to focus on the important stuff.

PHYSICAL CONTRADICTIONS AND SEPARATION PRINCIPLES

Many technical and nontechnical problems face Physical Contradictions: an object or a process has to have opposite characteristics to satisfy different requirements. Something has to be hot and cold, large and small, soft and hard, time-consuming and instantaneous, present and absent at the same time. A majority of solutions propose a compromise, a trade-off, or a narrow optimum. TRIZ accepts no compromises. For example, if something has to be hot and cold at the same time, it is not supposed to be lukewarm.

To solve Physical Contradictions, TRIZ uses four Separation principles. They are:

(1) Separation in Time

(2) Separation in Space

(3) Separation between Parts and the Whole (Parts of the system should have property A, but, as a whole, the system should not have property A.)

(4) Separation under Condition

As with other creativity tools, let us find inspiration in nature. To avoid self-pollination and to improve genetic diversity, some flowers develop stamens before or after the stigma. Thus the pollen can only reach the stigma from other flowers. This is Separation in Time.

Classical TRIZ postulates one always has to identify contradictions. I often use the four Separation principles by themselves.

A Physical Contradiction is a situation where an object or a process has to possess opposing characteristics at the same time. It "wants" to be hot and cold, long and short, heavy and light, bright and dark, smooth and rough, present and absent, etc. TRIZ also postulates that one should not accept compromises. As I said before, the typical solution is often a result of an optimization, a compromise. If something wants to be hot and cold, it ends up at an intermediate temperature, etc. Such solutions should be rejected, at least in the first pass.

Physical Contradictions and the four Separation principles can best be explained via examples.

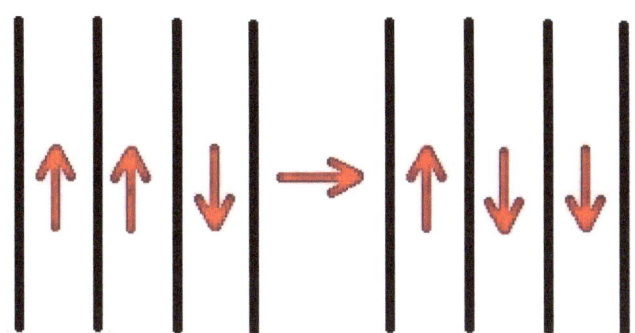

Fig. 18.1

Let us consider a busy highway (Fig. 18.1). One can observe that the traffic patterns change during the morning rush hour and the evening rush hour. Depending on the location of the residential versus business areas, a different number of cars may travel at a given direction in the morning versus in the evening. Applying the Separation in Time principle, one wants to allocate more lanes in the direction of prevailing traffic and to change the traffic direction on some lanes depending on the time of day. This is done on some bridges.

Here we applied the four Separation principles without a contradiction. If we were building a highway from scratch and were limited by the footprint or the budget, the contradiction may be wide (can handle more traffic) versus narrow (occupies less space, costs less). Separation in Time is often somewhat related to Dynamicity.

Separation in Space is somewhat related to Heterogeneity and Multiplication. Let us go back to the backcountry ski example (Fig. 5.3). The contradiction is that the skis are supposed to be smooth on the bottom to effortlessly slide on the snow and yet rough on the bottom to stay put even when the skier moves the other ski (rough versus smooth).

Let us apply Separation in Space. The front and rear sections of the ski are smooth. The middle section, the one that gets pressed to the ground by the skier's weight, is rough and bent upward. The act of pressing the middle section to the ground is also an example of Separation in Time (the ski moves versus remains stationary) and Under Condition (with applied weight). The optimized solution (a compromise) would be to have the bottom of the ski at some intermediate roughness, not too rough so it can still slide on the snow, yet not too smooth so that one can rest one's weight on it without slipping. This is obviously not what we want.

Here is another example of Separation in Space. My car has two climate zones, one for the driver and one for the passenger. The contradiction is hot (e.g., needed for a passenger dressed in light clothes but during winter weather) versus cold (e.g., driver dressed in a winter parka with the added heat in the car making him uncomfortable). The optimized solution (a compromise) would be to maintain the car at some intermediate temperature.

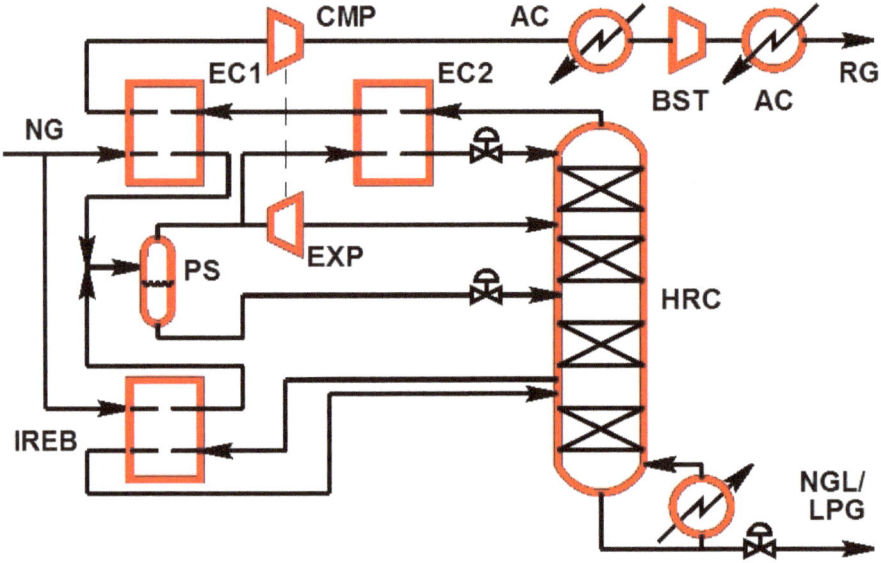

Fig. 18.2

Let us look at a more complex example. Fig. 18.2 shows a natural gas liquid/light petroleum gas (NGL/LPG) extraction system. It separates natural gas (NG) into Residue Gas (RG) and NGL/LPG: heavier components.

Without going into much detail, the system "wants" to be at high pressure so that the Residue Gas (RG) can be liquefied (liquefaction is more efficient at high pressure) or put into a pipeline (pressure drop losses are lower and pipes are smaller at high pressure). However, the "heavies" removal column (HRC) "wants" to be at low pressure (more efficient distillation or simply feasible distillation if the required high pressure is above the critical pressure). The contradiction is high pressure versus low pressure.

The third Separation principle is Separation between Parts and the Whole. The solution it dictates is as follows. The high-pressure natural gas feed (NG) is expanded in expander EXP and fed to the HRC. The column's gaseous product is then recompressed in compressors CMP and the BST (booster). CMP is driven by the expander EXP to recover part of the compression energy (Self-Service). That way the system's feed and principal product is at high pressure, but the column itself is at low pressure. For comparison, a typical optimized solution (a compromise) may involve running the column and liquefaction at the same pressure, as high as possible but low enough to achieve separation.

BIOMIMICRY

In several previous sections I mentioned that many idea-generation techniques can be traced to Mother Nature. Inventions are often a product of evolution in a certain discipline and so are nature's solutions. A classic example of an invention inspired by nature is Velcro, as evidenced by burs of certain plants sticking to our clothing.

The invention of a wheel is traced back to ancient Mesopotamia, but some cultures did not use it until recently. Yet wheels used for rapid travel are known in nature. For example, mother-of-pearl caterpillars curl up into a wheel shape to roll away from predators. Artificial irrigation was a breakthrough in food production in ancient Egypt. Bullfrogs dig trenches to supply water for their tadpoles. Certain plant-hopper insects use natural gears to help them jump.

The polar bear's fur is translucent. The sun's rays can penetrate it to be absorbed by the animal's dark skin. The fur minimizes the convective heat transfer. This principle can be applied to clothing articles that include, but are not limited to, jackets, parkas, pants, gloves, and hats.

I could give many more examples, but instead I refer the reader to my Literature section herein and the previous sections.

MONTE CARLO

Random-entry idea-generation technique ("stichomancy") relies on picking a random noun from a dictionary to generate ideas to solve a particular problem. At this point in writing this book, I picked up my Webster's dictionary, flipped it open, and pointed to a spot on the page. My finger chose "sitting duck." Let us write down the characteristics related to this noun. Duck: can swim, can fly, waterproof feathers, flat bill, migrates from place to place in search of food and better weather.

What if we apply it to the gas-extraction process in Fig. 18.2? We then have so-called FPSO: Floating Production Storage and Off-Loading, utilizing a plant mounted on shipboard that can move from one gas well to another.

The Monte Carlo method was developed by Nicholas Metropolis and Polish-born Stanislaw Ulam, first described in 1941 in the *Journal of the American Statistical Association*. It uses random numbers to obtain numerical results.

Fig. 19.1

I developed my own abstract painting technique that uses random numbers (Figs. 19.1 and 19.2). I fold an adhesive sticker and cut it with scissors to produce repetitive geometrical patterns. Then I transfer them to a canvas. Inside each pattern I assign zones. Then I use a random number generator to assign colors to each zone. Each color has a corresponding number: 1 = green, 2 = blue, etc. I generate numbers from 1 to 10 for each zone. That way the computer picks colors. I would never have thought about color combinations produced that way.

The title of my art piece is in the runic alphabet. What does it say? Hint: *x**x*x, where X is the repeating character, is also used in my initials.

Fig.19.2

LITERATURE

1) *40 Principles: TRIZ Keys to Innovation* by Genrich Altshuller

2) *And Suddenly the Inventor Appeared: TRIZ, the Theory of Inventive Problem Solving* by Genrich Altshuller

3) *Innovation Algorithm: TRIZ, Systematic Innovation and Technical Creativity* by Genrich Altshuller

4) *The Ideal Result: What It Is and How to Achieve It* by Jack Hipple

5) *TRIZ for Engineers: Enabling Inventive Problem Solving* by Karen Gadd

6) *Da Vinci and the 40 Answers* by Mark L. Fox

7) *Thinkertoys: A Handbook of Creative-Thinking Techniques* by Michael Michalko

8) *Inside the Box: A Proven System of Creativity for Breakthrough* by Drew Boyd and Jacob Goldenberg

9) *SCAMPER: Creative Games and Activities for Imagination Development* by Bob Eberle

10) *SCAMPER On: More Creative Games and Activities for Imagination Development* by Bob Eberle

11) *The Creativity Tools Memory Jogger* by Michael Brassard and Diane Ritter

12) *The Art of Innovation: Lessons in Creativity from IDEO, America's Leading Design Firm* by Tom Kelley, Jonathan Littman, and Tom Peters

13) *Cats' Paws and Catapults: Mechanical Worlds of Nature and People* by Steven Vogel and Kathryn K. Davis

98

14) *Weird Nature: An Astonishing Exploration of Nature's Strangest Behavior* by John Downer

15) *Kreativitätstechniken* by Matthias Nöllke

16) *Sex, Drugs, Einstein & Elves: Sushi, Psychedelics, Parallel Universes and the Quest for Transcendence* by Clifford A. Pickover

17) *How Would You Move Mount Fuji? Microsoft's Cult of the Puzzle—How the World's Smartest Companies Select the Most Creative Thinkers* by William Poundstone

18) *Are You Smart Enough to Work at Google? Trick Questions, Zen-like Riddles, Insanely Difficult Puzzles, and Other Devious Interviewing Techniques You ... Know to Get a Job Anywhere in the New Economy* by William Poundstone

19) *How to Ace the Brainteaser Interview*, by John Kador

20) *Nature Got There First: Inventions Inspired by Nature* by Phil Gates

21) *Ingenious Women: From Tincture to Saffron to Flying Machines* by Deborah Jaffe

22) *Sneaky Uses of Everyday Things* by Cy Tymony

23) *Implementing TRIZ at Air Products*, *KM Review*, volume 9, issue 3 by Adam Brostow

24) *Ancient Inventions* by Peter James and Nick Thorpe

25) *Technology in the Ancient World* by Henry Hodges and Judith Newcomer

26) *1001 Inventions That Changed the World* by J. Challoner